NORTH BY NORTHWESTERN

NORTH BY NORTHWESTERN

A Seafaring Family on Deadly Alaskan Waters

Captain Sig Hansen
and
Mark Sundeen

SIMON &
SCHUSTER

London · New York · Sydney · Toronto

A CBS COMPANY

First published in Great Britain in 2010 by Simon & Schuster UK Ltd
A CBS COMPANY

NORTH TO ALASKA
Words and Music by JOHNNY HORTON and TILLMAN B. FRANKS
© 1960 TWENTIETH CENTURY MUSIC CORPORATION
© Renewed 1988 EMI ROBBINS CATALOG INC.
All Rights Controlled by EMI ROBBINS CATALOG INC. (Publishing)
and ALFRED PUBLISHING CO., INC. (Print)
All Rights Reserved   Used by Permission

Sink the Bismarck
Words and Music by Johnny Horton and Tillman Franks
Copyright © 1960 UNIVERSAL–CEDARWOOD PUBLISHING
Copyright Renewed
All Rights Reserved   Used by Permission
Reprinted by permission of Hal Leonard Corporation

1 3 5 7 9 10 8 6 4 2

Simon & Schuster UK Ltd
1st Floor
222 Gray's Inn Road
London
WC1X 8HB

www.simonandschuster.co.uk

Simon & Schuster Australia
Sydney

A CIP catalogue copy for this book is available
from the British Library.

ISBN: 978-1-84737-889-7 (hardback)
ISBN: 978-1-84737-884-2 (trade paperback)

Designed by Phil Mazzone
Map by Paul J. Pugliesi

Printed in the UK by CPI Mackays, Chatham ME5 8TD

My brothers and I dedicate this book
to our father, Sverre Hansen,
our uncle, Karl Johan Hansen,
and all the men who pioneered the crab industry
and paved the way for us to follow.

CONTENTS

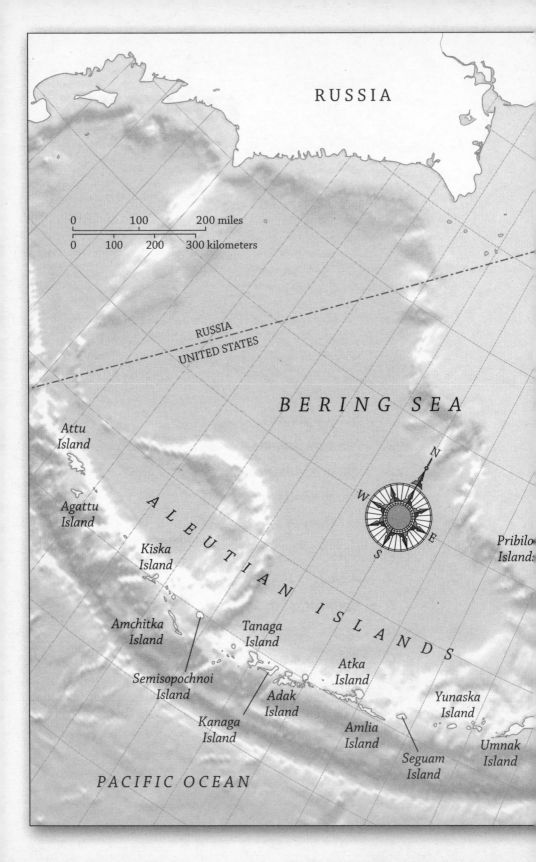

NORTH *BY* NORTHWESTERN

Prologue

When it started, the captain was in his bunk. It was eight in the morning, in early December. The gray dawn seeped through the black Alaskan night. Fueled by cigarettes and coffee, the captain was a light sleeper. But this night had been worse than usual. The vessel was bucking in 20-foot seas, the arctic wind howling at 60 knots. He had been lying half-awake for four hours, the heaves of the boat knocking him against the wall.

The captain and his three-man crew were jogging into the seas northeast of Dutch Harbor, in the Bering Sea, on the way to pull the crab pots they had set days before. The captain had manned the wheel the previous night, and finally at three in the morning he'd let the deckhands take turns on watch while he tried to catch some sleep in his stateroom. No luck. Now it was almost a relief to hear the footsteps clomping in from the wheelhouse. It meant something was wrong, but at least it was an excuse to get up. The door swung open.

"I'm losing power," said Krist. "The steering, the radios, the whole works."

The captain threw his feet off his narrow bunk. He had grown up in Norway during the war, and when his village was occupied by Nazis he admired the tough British who vowed never to surrender. One of his life heroes was the steely prime minister, Winston Churchill. The captain pulled on his slippers and stood. He wore heavy trousers and a sweater. The captain always slept fully clothed. He followed Krist out the door.

Of course the captain wasn't happy that the power was failing, but he was relieved to be getting out of bed. He got bored and restless when he wasn't working. He liked to work. He liked to work long hours, twenty hours a day, or sometimes when the fishing was good, he didn't bother sleeping at all. That's the Lutheran ethic his father had drilled into him. Laziness was what ailed the weak and the cowardly. Work was good for you, the harder the better. That's what he was here for.

As Krist returned to the captain's chair and lay a hand on the throttle, the captain steadied himself in the doorway and lit a Pall Mall. The boat rose in the waves and collided with the crests, then dropped weightlessly into the trough. The seas were rough, but the captain had seen much worse. The fact that he could walk around the wheelhouse without holding on to anything, without being knocked off his feet, told him the seas were nothing to worry about.

He stood beside Krist at the console. The other two deckhands were asleep downstairs. Off to the west the sky was silky black. It was hours before sunrise, nothing to see out there anyway, just hundreds of miles of cold ocean between here and Russia. To the southeast, the jagged silhouette of sea cliffs came into focus on the horizon. The boat was only a few miles from Akun Island. If worse came to worst they could steam toward land and find a sheltered cove and make whatever repairs were required. Among the captain and the crew, there were mechanics, welders, painters, carpenters, and even firefighters. They could fix just about anything.

There were all sorts of reasons a boat could lose her steering out at sea. A hydraulic hose could have ruptured and drained the fluids that build pressure to move the rudder. A cable could have snapped. A line could be wrapped around the propeller—maybe they'd run over the buoy of some other crabber's derelict pot—although this didn't seem likely because the hum of the engine down below sounded normal. That was a good sign, but losing all power was another problem altogether. Maybe it was something simple, like a short in the electrical system, or a breaker that flipped. The captain was not alarmed.

"I'll go take a look," he told Krist.

The captain took a drag on his cigarette and pushed through the door of the wheelhouse to the upper deck. He was met by a blast of arctic cold as he emerged into a frozen nightmare. Icicles clung to the eaves of the wheelhouse, and a crust of ice was caked on the rails. The boat rocked and pitched. He placed his foot carefully now, the iced deck cold beneath his thin soles. Out in the dark gray morning he saw dim flashes of whitecaps, but mostly the monstrous waves were invisible, they were ghosts. The captain braced against the mast and looked down to survey the deck. They had already dropped all the crab pots, and yesterday he had ordered the men to scrub the planks, so the deck was clear except for a white blanket of snow from amidship to stern. Spools of line were coiled neatly, encased in thick frost. The crane creaked overhead like a gallows. Life rings hung on their hooks, frozen solid to the wall behind them so that not even the arctic gales disturbed them. As soon as he solved this power problem the captain would wake the crew and get them out here chopping ice. The ice weighed tons and rendered the boat so top-heavy it might capsize. The wind howled in the captain's ears and he took a cold lungful of the sea air along with the hot tobacco smoke. The wind pierced his sweater. He shivered.

The captain had started fishing at age fourteen, with his father, who had started fishing as a boy with his own father. Like the rest of his crew, the captain came from the island of Karmoy, a fishing port on Norway's western shore. Back then, young men did not have a lot of options. If you couldn't get into school, and your family didn't own farming land, you went to sea.

The captain had gone to sea. He fished for herring and cod, up and down the Norway coast, as far away as Iceland. There were a few good years of herring, but then it was overfished. The boats lay in port. Times were hard. When he was twenty, the captain lit out for America. He fetched port in Seattle, in the old Norwegian waterfront neighborhood of Ballard. He was young and strong and could work hard and he spoke a few words of English. He did what fishermen have done for hundreds of years: walked the docks for work. He was hired on a trawler, the *Western Flyer*, dragging for bottom fish. He never looked back.

Those days were far behind him now. The fishing around Seattle had declined, and the captain joined an intrepid band of fishermen to explore the northern waters of Alaska. What they found was a gold rush—the red spiny kind that walked on eight legs. King crab. There was a fortune to be made in the Bering Sea, but the risks were high: brutal storms with 60-foot waves, hurricane winds of 130 knots, freezing spray, and treacherous ice caked on every surface. Older fishermen thought it was suicidal.

In those days Dutch Harbor—which would eventually surpass San Diego as the richest commercial fishing port in the United States— was just a weather-beaten outpost with a handful of natives and soldiers, and a very rough bar to drink in. Japanese vessels had trawled for crab for decades, but it wasn't until the daring Americans and Norwegians began tinkering with boxlike steel traps that crab fishing reaped huge profits. The captain and his ilk were pioneering crab fishing on the fly.

It was damn hard work, and dangerous. Men heaved the 700-pound pots around the deck with their hands, at the mercy of waves that could sweep them right off deck, or even knock a boat upside down. The Bering Sea was a place of hideous extremes. There were a dozen ways to die out here, and the captain knew men who had succumbed to all of them. His own uncle had been knocked overboard when a steel cable snapped and slapped him like a bullwhip on the head. They never found his body.

The captain had seen men washed overboard. Some brought it upon themselves by reaching too far over the rail to haul in a line, by scaling the stack of pots and losing a grip with their frozen fingers, or by stepping on the wrong side of a pot as it swung from a crane and getting clubbed into the froth. Then there were the men who did everything right, and were still washed overboard by 50-foot rogue waves that swept across the deck and uprooted steel machinery by the bolts and tossed 7-foot crab pots like toy blocks. It was just luck, or fate. Some of the bravest and most cautious men died, while some of the lazy men, and the cowards, were given another chance.

Once submerged in the Bering Sea, a person has about four minutes before he dies. The 36-degree water triggers a gulp reflex: Some men inhale water instantly and sink to the bottom. The panicking human heart pounds like a jackhammer—some men succumb to heart attacks. Those strong enough to last four minutes lose the ability to kick their legs and tread water, and then, with what survivors fished out in the nick of time describe as a peaceful feeling of acceptance, they stop struggling and let the sea pull them below.

There was a chance that in those four minutes a man overboard could be rescued, but it wasn't likely. It could take two minutes just to turn the boat around. If the currents worked in the man's favor, he might drift alongside the boat, and a deckhand with steady nerves and a good arm could throw him a life ring. The drowning man—if his palsied fingers could make a fist—might grab hold, and be hauled to safety. More often the crew couldn't even see the lost man, his head a floating speck among monster waves, frothing whitecaps, and blinding wind.

Then there were all the ways a boat could sink. A hatch could be left unsecured and the sea would pour into the storage chamber in the stern—the lazarette. A seal might blow and the engine room would flood. In rough seas the captain might not notice until the stern was submerged. Maybe the bilge pumps would work, or maybe they wouldn't. The boat might list to one side and finally lie down, or she might bob vertically while the crew clung desperately to her exposed snout. Sometimes, a fire might start belowdecks, usually caused by an electrical short, and within minutes the engine would be disabled and the power would go. Without bilge pumps and a motor, a swamped boat would quickly sink. Without a functioning radio the crew wouldn't get off a mayday. Some boats had no mechanical failures, but were simply caught off guard by a rogue wave as tall as a ten-story building that would smash through the windows, flood the cabin, and capsize the boat. If the men were belowdecks they would never see daylight again, dropping to the ocean floor in a steel coffin.

Then there were the boats that simply disappeared—no maydays, no witnesses, no flotsam, no survivors. One day they were there, the next they were gone.

Sometimes the crew could escape a sinking ship to a rubber life raft, but this was not the end of the emergency—it was the beginning of another ordeal with a variety of likely bad endings. Momentarily safe from doom, the men might simply freeze to death, as they drifted for days with no rescuers in sight. There were no radios on life rafts. They might fire a flare into the night sky, but who was out here to see it? A wave might capsize the raft, and the men who got separated had just those dire four minutes. Some rafts might drift toward land, and as the men optimistically leapt in the water and swam for shore, they were smashed up in the twenty-foot breakers. Their lifeless bodies would wash up on the cliffs with the kelp. Others might land safely, only to find themselves stranded in the bitter Alaskan wilderness, starving, unable to build a fire, and hunted by grizzly bears.

None of the options were good. There was no safe place on the Bering Sea. The only place the captain wanted to be was on his ship. His wife and children were back in Seattle, expecting his safe return. He needed to be careful. The captain was not a regular churchgoer—spending ten months a year at sea didn't allow it—but his Lutheran upbringing haunted him, and like fishermen before, he prayed for safe passage. The old saying that there are no atheists in foxholes—same thing goes for Bering Sea crab boats.

The captain gripped the wooden rail. Everything on deck looked normal. Then he caught a whiff of something odd: the oily black smoke seeping from the smokestack and whipping in the wind—it didn't smell like diesel exhaust. It was hard to explain. Smokier, maybe. The captain rushed back into the wheelhouse and lowered himself down the ladder to the galley. He approached the door leading belowdecks to the engine room. He laid his hand on the latch. It was hot. Something was wrong. Smoke seeped from the cowling. The captain pushed open the door and there, in the bowels of this wooden vessel, flames leapt from the machinery and lapped against the walls. A blast of heat and smoke pushed him back, and the captain slammed the door and turned and ran to wake the crew.

"Fire!" he yelled. "All hands on deck!"

My name is Sig Hansen, crab fisherman, captain of the fishing vessel *Northwestern*. The captain who awoke that December morning to find the engine room ablaze was my father, Sverre (pronounced *Svare-ee*) Hansen, and the boat that caught fire was the F/V *Foremost*. Until my brothers and I sat down to write our family's history, I had never heard the whole story of the sinking of the *Foremost*. The story was passed around for years between my father's friends, whom I have fished with all my life, but it never reached me until now. Of the four men aboard the *Foremost* that morning, only one is still alive.

I have been a fisherman all my life. I began when I was twelve, and have done it for more than thirty years. It is the only work I know. When my father died, my brothers and I became owners of his ship, the *Northwestern*, a 125-foot steel crabber built to withstand the Bering Sea in winter. I am the captain, and my brothers Norman and Edgar alternate between engineer and deck boss. This is the life I hoped for— nothing more, nothing less. When I decided as a boy that I'd be a fishermen like my father, the last thing I ever imagined it would bring me was notoriety.

A few years ago, however, my brothers and I decided to allow a film crew aboard the *Northwestern*. Nobody could ever have predicted what it would lead to. Almost overnight we went from chopping bait in the freezing sleet to signing autographs at Disney World. Through the popularity of Discovery Channel's *Deadliest Catch*, Alaska crab fishing has gone from a deliberately secret society to a worldwide phenomenon, with millions of fans and viewers in 150 countries. We've been on talk shows. We designed an Xbox video game to enable people to go crab fishing (in their living rooms). Folks can also buy all sorts of things—emblazoned with the Hansen name and the *Northwestern* logo—from our very own Rogue beer to fish sticks to rain jackets to four types of Crab Louie sauce. To think that complete strangers approach me in airports or malls and want to know what kind of bait I'm using, or how many king crab I'm averaging per pot—it astounds me. The other day a lady came up to Edgar down at the Seattle docks and said that when she asked her son what he wanted to be for Halloween, the kid said, "Edgar." I wish Dad were around to see it.

However, if all you know about commercial fishing is what you've seen on the *Deadliest Catch,* you've only seen the tip of the iceberg. The media portrays my brothers and me and the other captains and crew as the ultimate tough guys. To which I say: You ain't seen nothing. You should have seen my dad. You should have seen my granddad, my Uncle Karl, and all the men who came over with them from Norway or ventured north from Seattle to pioneer the crab industry, long before cable television, GPS, satellite phones, and computer depth-finders and plotters. Hell, they were doing it in wooden boats.

So when I decided to write a book, I knew I wanted to tell about more than myself. I wanted to tell the story of my family and all the men who built this industry. I'm just a speck in the history of fishermen and sailors, which dates back generations and centuries. To understand how I came to sit in the wheelhouse of the *Northwestern,* you have to understand the heritage of Norse fishermen, the waves of immigration that brought them to America, and the wild courage of the first men with the nerve to search Alaskan waters for the prized crabs. Most of all, you need to understand my father, Sverre Hansen. For my brothers Norman and Edgar, as well as me, each day at sea is spent trying to prove ourselves worthy of the Hansen name and trying to live up to the example that our father set.

My father was very proud of his roots. When we were small boys he would sit us in his lap and say, "Always be proud of your heritage. You have to keep it going: what you are, who you are, where you come from." That's why I'm writing this book.

The Hansen family descends from the Vikings who roamed and ruled the northern seas for centuries. In those times, in the cold lands of Norway, Sweden, Denmark, and Iceland, the ocean was the frontier, and the legends grew from the sea, from the brave sailors who set out for riches and adventure. In Viking lore, those legends—the sagas—were passed from father to son, from generation to generation.

My brothers and I learned to fish from our father when we were boys. He learned from his father, who learned from his father. I am proud that the life I have chosen is to work hard, to face the dangers of the seas, and to pull a living from the ocean, just as my father did, and those before him. The Hansens are a family of fishermen, of sailors, of captains.

This is our saga.

1

SON OF NORWAY

My grandfather had a scar that ran all the way down his leg from hip to ankle. He used to come over from the Old Country to fish in Alaska. When I was a kid he was too old to do much hard work, but he liked to come anyway, just for fun; an old-timer who wanted to be with his family. He had been fishing herring and cod since he was a boy.

My grandfather had the accident on a supply ship on the North Sea. While they were putting a hatch cover down he got his leg caught in the opening, and this giant steel lid—the hatch—fell on it, and split his leg from top to bottom.

When we visited Norway that year, he was on crutches, walking back and forth in the living room trying to recover. The stitches were still in. It was a gruesome wound, like something you'd see on Frankenstein. He really got mangled. My brothers and I kept staring. Of course it scared us, but it wasn't so traumatic that we decided crab fishing was too dangerous. We knew it was part of the job. He just got unlucky.

His name was Sigurd Hansen, and that's who I was named after. My other grandfather's name was Jakob, and when I was born, each grandfather thought I should be his namesake. My parents debated. In Norway, even if you didn't choose a relative's exact name, it was considered an honor to pick one that began with the same letter. So my parents settled on Sigurd Johnny Hansen—with the middle name pronounced Yonny. Even though it's an honor to be named after my grandfathers, I can tell you that when you grow up in America in the 1970s with a name like Sigurd Johnny, you're going to get a lot of black eyes.

I was born in 1966, in Ballard, the Scandinavian part of Seattle, by the ship canal. All my parents' friends were Norwegians, and fisherman. There were a few Swedes and Danes, but other than that we didn't really socialize with anyone else. My parents didn't have to speak English much. They didn't go to PTA meetings or things like that. My mother's English wasn't great. My father was gone nine months out of the year fishing, so my mother was home with us kids and didn't get out to the broader public much. All she needed to know in English was how to get by at the grocery store—how much things cost and how to pay for them. In first grade they sent me back home with a note that read: TEACH HIM ENGLISH. STOP SPEAKING NORWEGIAN.

Right about the time I was born my parents moved out of Ballard to the northern part of Seattle to a bigger house with a yard. A lot of the Norwegians of my parents' age were moving north, too. This small community was a great place to grow up. When my brothers and I walked to junior high school we'd stop by our cousins' house and walk the final blocks together. We went to the Rock of Ages Lutheran Church in Ballard. In elementary school and Sunday school there were a lot of Norwegian kids, and we spoke the language to each other. Sometimes we'd ride our bikes all the way down to Fishermen's Terminal in Ballard to check out the boats. Our fathers were all fishermen. A lot of us would go back to Norway for the summer or Christmas break, and run into each other back there.

Just like normal American kids, my brothers and I were on soccer and baseball teams, played in the school marching band, and sometimes even babysat to earn spending money. We also knew we were different. We knew we were Norwegian, and we knew we were fishermen. Mom says she always knew I would become a fisherman. In school while the other kids learned A-B-C and 1-2-3, I would draw a boat and a crab pot and a black swirl coming out of the smokestack.

I got my first chance to see my father at work in Alaska in 1978 when I was twelve. I rode the *Northwestern* with him from Seattle across the Gulf of Alaska to the Aleutian Island chain. It was a one thousand seven hundred-mile trip that took more than a week, and most of the time I was terribly seasick. From there we motored another four hundred miles north through the Bering Sea, three days without sight of land. We finally dropped anchor off St. Matthew Island—a deserted outcropping in the far northern Bering Sea, right by Russia. Bizarre formations of volcanic rock rose up from the shoreline. Since it was summer near the Arctic Circle, the sun only dipped below the horizon for an hour or so and the skies never darkened. I was wide-eyed with wonder. Suddenly my world of the Seattle suburbs had expanded into something vast and strange and full of adventure. My dad's life as an Alaskan fisherman—to me an abstract concept that he'd talked about for years—was suddenly vivid and real.

We fished blue crab that year, but I was just a kid and not much use. What seared a stronger impression on me than the actual fishing was the thrill of exploring this exotic new world. At one point the fishermen went on strike and we were laid up in harbor with 150 other boats. To pass the time, a few of the other kids and I took a skiff and ventured onto the island. We found deer darting across meadows of wildflowers and rams roaming the rocky hillsides. We discovered a stream so thick with Dolly Varden trout we were able to scoop them up with our bare hands. We wanted to fish the lake that the stream poured into, so we returned to the ship and crafted a reel by wrapping dental floss around an old strawberry can. We bent nails into hooks and sharpened them in the grinder and splayed bits of polyfiber line into lures. We returned to our lake and tossed out our lines with a bit of lead for weight and pulled in those Dolly Vardens one after the other. It was the height of summer and the weather was warm.

Another thrill of that summer was to finally see in action all those beautiful crab boats that I'd been drawing pictures of for years. My friends and I took the skiff from boat to boat and climbed aboard to meet the crews and have a look. A lot of the guys were friends of my father; if they weren't, they were still friendly and welcomed us aboard to check out the wheelhouse and the deck gear. We pored over every inch, as if we were in a museum. I remember a crabber called the *Neptune*. That big wheelhouse, all painted black and white, looked like a race car. What a beautiful boat. By the time I was fifteen I knew every boat in the fleet. I'd see them on the horizon, and I could pick them out just like that. The older guys on the crew would just look at me, shake their heads, and say, "How the hell can you see that far?"

Probably the most important part of that summer was getting to work alongside my father and the other men who would become my lifelong mentors. They were larger than life. Dad's friend Oddvar Medhaug worked on deck that summer. He was a classic Norwegian—hardworking and stubborn, from the same mold and same town as my dad. In later years he would become one of the most successful skippers in the Alaskan fleet—what they called a highliner—but that summer he was still a deckhand. I idolized him. When he came down from the deck to wash dishes in the galley, I snapped pictures of him, as if he were a celebrity.

After two months, I had to get back home to Seattle for junior high school. They dropped me off on St. Paul Island, another rock in the Bering Sea. I had to overnight by myself. The bar was next door to the hotel. The natives and fishermen were drinking and it started to get pretty wild and very loud. I could hear bottles crashing and fights breaking out. Only twelve years old, I was terrified and hid beneath the window in my room. In the morning I flew to Cold Bay, which is a tiny outpost at the tip of the Alaskan Peninsula where the Aleutians start. There was nothing but an airstrip and a store and a few weather-beaten buildings. Then Anchorage, then Seattle. It took a couple of days.

The summer of 1978 was a huge adventure filled with hardship and fear and wonder and excitement that left me wanting more. I had dis-covered my ambition: I wanted to be a fisherman—and a man—like my father. From that point forward, my life became a quest to prove myself a real fisherman in his eyes.

The next summer I went to Norway and fished with a third uncle named Hans. We fished for cash, and he'd pay me some spending money under the table. Once I turned fourteen, in 1980, and was con-firmed in the church, I felt like I was an adult, and was ready to get out of Seattle and find adventurous work. So I left school early that sum-mer and went gillnetting for red salmon up in Bristol Bay with John Jakobsen. He was a friend of my dad's from Karmoy—a hell of a great fisherman and a mentor. John didn't hire many greenhorns, but he made an exception for me as a favor to my dad, and because I already knew my way around a boat.

Before I left, my father took me down to the supply shop and bought me boots, raingear, and a duffel bag. He even tried to buy me one of those floppy orange southwestern hats, like the old fishermen wear, but there I drew the line. As much as I wanted to emulate him, I was just too young to dress like an old-timer. The Old Man even helped me pack my seabag. He paced around the house, and double-checked my gear. I could tell he was nervous about me going up to Alaska without him, but he never said a word about it. He wasn't the type of guy who easily expressed his emotions. Instead, he gave me the type of fatherly advice one gets from a stoic old Norwegian fisher-men, "Keep your mouth shut, do what he tells you, and everything will be fine." Then he put me on a plane and I flew north.

Bristol Bay is the southeast corner of the Bering Sea, formed where the Aleutian Peninsula juts out from the Alaskan mainland, a few hundred miles west of Anchorage. Fed by a number of coldwater rivers, Bristol Bay is the richest salmon ground in the world.

John's boat was the *Jennifer B,* a thirty-two-foot aluminum boat with a blue hull and a white deck, and old tires lashed to the rails like bumpers, so it could bob against docks and other boats without get-ting damaged. Most of the boat was deck, with a tiny wheelhouse above and four bunks below. The entire crew consisted of John, another older man named Bjarne Sjoen, and me. We spoke mostly Norwegian while we worked.

The minute the season opened, Bristol Bay was chaos, with hun-dreds of boats competing for the millions of salmon making their run. The whole season lasted only five or six weeks. We fished for short spells, depending on when the Alaska Department of Fish and Game called an opening, which lasted twelve, twenty-four, or thirty-six hours.

Like the rest of the fleet, the *Jennifer B* was a drift netter. A drift net is a huge, flat rectangle made of mesh—450 feet long and 10 feet tall—that hangs in the water like a curtain. Picture a gigantic tennis

My first real job on a salmon boat, the *Jennifer B*, in Bristol Bay, 1980. *(Courtesy of the Hansen Family)*

net. We slowly unfurled the thing from a drum on the stern, and once it was in the water, the top edge—called the cork line—floated on the surface while the bottom edge—the lead line—sank below. Fish can't see the mesh underwater. Their heads pass through the holes in the mesh, but their bodies are too big. Their gills get stuck, and they are caught.

We set three nets, let them soak, and then hauled them aboard. While John Jakobsen manned the wheel, Bjarne and I hauled the net and stacked it on deck, trying to keep it from tangling. If the fishing was good, we'd find three hundred salmon in each net. If it was heavy fishing, we could find as many as a thousand. The real work began when the net—teeming with flopping fish—was stacked on deck. Bjarne and I started fish picking—rifling through the nine thousand square feet of net and pulling the fish. We picked as fast as possible without damaging the salmon. The faster we picked, the more fish we caught, and the more money we made.

In addition to my father's admonition that I keep my mouth shut and follow orders, I'd been given two very wise bits of advice that I've carried with me ever since. First, one of my dad's friends had said, Never let the other guy beat you. So I worked myself ragged. It didn't matter that I was just fourteen—I picked those fish as fast or faster than the grown men. Second, my grandfather's advice was to get as much sleep as I could, whenever I could, since I'd never know when I'd sleep next. So I did. Captain John would tease me because if we had a twenty-minute run I'd hurry down to my bunk and close my eyes. Once I even tried to sleep in my raingear.

For a kid like me trying to become a real fishermen, the most memorable part of Bristol Bay was the nights at the docks, hanging around with the guys I idolized. If the tide was high enough, we'd tie up at the docks, and all the old-timers would get together and tell stories. Along with John Jakobsen, a lot of my dad's friends from Karmoy were there, like Oddvar Medhaug and John Johannessen. I wasn't the only young kid. John Johannessen's sons, Lloyd and Norman, who were a few years older than me, worked up there as well. Eventually my brother Norman worked a few seasons. So we had a multigenerational Norwegian community, and what we kids wanted most of all was to be men like our dads.

While the old guys were sitting up in the camp, the kids were hanging out on the boats. Now and then we'd steal a couple of beers from their cooler. We'd sit around the wheelhouse trying to act grown up—sipping beers, smoking cigarettes, and looking at *Playboy*. One night John Johannessen walked in. Because I was a greenhorn—and because they knew my father—the older guys hazed me, a tradition probably as old as fishing. Johannessen took a sidelong glance at my beer. He joined in the conversation as if he hadn't noticed what we were up to. "I used to chase your dad home from school," he said. "He was just a little runt. Everyday I'd beat him up."

Then when he thought I wasn't looking, he stealthily tapped his cigarette and dropped the ash into my can of beer. He was as smooth as a magician, his sleight of hand so subtle that I thought I'd imagined it. There was no visible evidence on the can rim. I kept watching out of the corner of my eye, and sure enough he did it again, using my precious pilfered beer as an ashtray. So now I had a dilemma: Do I complain that he ruined my beer, and then have to explain what I was doing drinking it in the first place? Or do I just do what I imagined a real man would do, and drink up? Holding my breath to not taste it, I picked up the can and chugged down the warm suds, the cigarette ash scratching my throat on the way down.

At the end of the season, I got my first official paycheck. It was a couple thousand dollars—more money than I'd ever seen. I didn't even know what to do with it. My parents helped me set up a bank account, and I decided to save it all. Even at that age, I was frugal. Then Dad explained to me how income tax worked. He said there were two options: I could skip from boat to boat, and avoid filing returns, and maybe they'd never catch up with me. Or if I wanted to be a part of the system, a real member of the fishing industry, I could file with the IRS and start paying taxes. Of course I wanted to do what he did. So at age fourteen I cut my first check to Uncle Sam. It sucked.

I fished with John Jakobsen for half of four summers, all through high school. I was a hard worker and it wasn't long before I was being paid the same as the guys twice my age. As soon as spring arrived I'd be hanging out at John's house, around the corner from my family's house in Seattle, shooting pool, talking about fishing, just itching to once more get on a plane for Alaska. After salmon season I'd get off that boat and go to Norway to fish mackerel and cod.

Where I really wanted to be was on the Bering Sea. Salmon fishing was fun, and the money was good, but in my mind, crab fishing was the major league. That's where the big boats were, that's where the captains were making big money, and that's where my father was.

When I was fifteen, my father hired me as a greenhorn deckhand on the *Northwestern,* and I returned to St. Matthew. I worked elbow to elbow with guys I idolized and felt like I had finally arrived. Fritjoff Peterson was a few years older than me and had been working on deck on the *Northwestern* since he was sixteen. He was huge, about six and a half feet tall, and he went to the public school in Ballard when most of the Norwegians had moved up to the suburbs. He was born in Norway, and one of the only European kids in the school. Being that big, being named Fritjoff, and speaking with a funny accent made him a target. Everyone wanted to fight him. So his parents were glad to ship him up north as soon as possible where he fit right in. Pretty soon he bought a Corvette—I guess that meant he was making cash like a real crab fisherman.

Another one of my mentors on the *Northwestern* was Mangor Ferkingstad, a Norwegian who was born on the East Coast but grew up in Karmoy fishing on the North Sea. When he was twenty he came to Alaska and started working for my dad. Like a lot of guys from the Old Country, Mangor loved American muscle cars. With his earnings he bought a souped-up old Cougar. Later Dad lent him money to buy a Monte Carlo that he still has today. To this day Mangor and I are like brothers.

Nineteen eighty-one was another great summer—a real adventure. In between crab seasons, Mangor, Fritjoff, Brad Parker, the engineer,

(Left to right) I, Fritjoff Peterson, Mangor Ferkingstad, and my brother Norman: the young crew of the *Northwestern,* circa 1982. *(Courtesy of the Hansen Family)*

and I took the skiff and went beachcombing on the tiny unpopulated islands. On one such trip we returned from our hike to find the skiff had floated away. The *Northwestern* was anchored five miles out, and we hadn't brought a radio. Luckily, Brad saved the day. He kicked off his boots and swam after it, climbed aboard, and brought it to shore. By then he was shivering cold, so before we could make the crossing back to the ship, we built a fire and dried his clothes around it.

On one such exploration on St. Matthew Island, Mangor and Fritjoff and I saw an antenna on a hillside, and hiked up to investigate. We came across a tent, and inside it was an emaciated, bearded man—who looked like Robinson Crusoe—surrounded by a million dollars of electronic equipment. He nearly flipped when he saw us. "Can you help me get off this island?" he asked. He had been hired by one of the oil companies to spend the summer monitoring the shipping lanes but was having trouble: The helicopter that was supposed to pick him up couldn't land on the island because it was a bird sanctuary; the boat that had come to fetch him didn't have a skiff that could get to shore; and he'd become too weak to hike to the far side of the island where his food drops were made. When we arrived he was down to apples and water. The guy's name was Matt, and he was from Houston, so we called him Matt Houston. We offered to take him back to the *Northwestern*. He was relieved. It was clear he hadn't had any company for months, and that it was getting to him.

"What do you do for fun out here?" I asked.

"Chase foxes around," he muttered.

We brought Houston onboard and sat him in the galley and began to feed him. Matt Houston ate more than any man I'd ever seen—plate after plate with no pause. Pork chops, potatoes, spaghetti—he inhaled whatever we served. We finally contacted his ship and delivered him safely.

Everywhere we went we'd find ourselves in situations that seemed to come straight out of a storybook. Once, after fishing for red crab near Nome, we anchored and went into town. It felt like the wild frontier. The deckhands snuck me into a saloon, telling me to duck as we passed the bartender, then led me into a dark booth. Outside, tourists paraded up and down the street, which had been dug up to lay a sewer line. One tourist started poking around in the rubble and unearthed a gold nugget the size of his hand.

The guys I met in those days were real characters—in good ways and bad—and the wildness of the world became more apparent to me with each season. The native salmon fishermen in Bristol Bay had a superstition: If they could convince a women to urinate on the nets, they would have good luck. Well, the old Norwegians decided they wanted some good luck, too. One night I awoke in my bunk to the sounds of shouts and laughter. Then I heard a woman's voice. I peeked abovedeck and saw that they'd brought a gal back to the boat. They helped her climb aboard and up the pile of salmon nets. *What the hell was going on?* I blinked to make sure I was seeing what I thought I was seeing. Sure enough, she did her business on the nets, while the fishermen hooted and laughed. I sneaked back to my bunk. The superstition seemed to work—the next day we went out and caught so much salmon that we almost sank the boat.

Gradually the younger generation was handed more responsibility. After the salmon season in Bristol Bay, we worked crab seasons out of Dutch Harbor. Between seasons, the men wanted to fly home. Tickets were too expensive for us to go with them, so they left us behind to watch over the boats for a couple of weeks. They would tie up in Dutch Harbor, one of the only places in the Bering Sea to catch a plane to the mainland. Usually, we accompanied them to the airport, a tarmac along the sea cliffs riddled with potholes. There was no set schedule of flights. We waited around all day until the little twin-prop arrived—or didn't. A tinny whine echoing across the harbor signaled an approaching plane. Once it landed, the plane rolled up to the shack of a terminal. Luggage was unloaded and heaved into a mud puddle. Then our fathers boarded the plane. Even though I wasn't actually working Dutch Harbor king crab season—it occurred in the fall when I had to be in school—just being there made me feel like I'd been promoted.

If the Bering Sea is the major leagues of the fishing industry, then Dutch Harbor is Yankee Stadium—the place with the biggest boats, the biggest processors, the biggest paychecks, and the biggest egos. Dutch Harbor is formed where two islands almost touch. The big island is Unalaska, home to the village of the same name. The waterfront is lined by old houses, the Russian church and graveyard, and the Elbow

Room and Carl's Hotel—both of which are now closed. A salmon creek flows through town. The little island is Amaknak, which we usually call Dutch Harbor. In Amaknak you'll find the airport, the UniSea cannery, and the spit of land that forms the small bay where most boats tie up. Back then the town was a lot smaller, there were no buildings out on the spit, and the bridge connecting the two islands hadn't been completed, so you had to take a skiff back and forth, or hire a water taxi.

With our fathers gone, we would get around on a skiff. If our boat didn't have one, we'd borrow (steal) one and cruise around the harbor. We'd give people a ride, and they paid us in beer—a beer taxi—but if they missed the water taxi after the bar closed, they were out of luck. Some people tried to swim across the bay in survival suits, if they could find one. I heard that one night Fritjoff and Lloyd missed the taxi, so they took a couple of plastic totes from the cannery and paddled them across the harbor. As it turned out the totes had holes in them. "Fuck, we're sinking!" they yelled as they slipped below the freezing cold water. Fritjoff ended up swimming back to shore, but Lloyd kept on going, and paddled all the way across.

Those were such great summers to be a kid, all freedom and adventure. I alternated between salmon season in Bristol Bay and crab season in the Bering Sea, with an occasional trip to Norway. I even got to work a season with my Uncle Karl on a leaky old wooden boat he'd bought just for the permit that came with it. The other deckhand was my good friend Glenn Tony Pedersen, and we'd wake up in our bunks with our pillows and sleeping bags soaked like sponges. Often we'd roll out of our bunks and find ourselves shin-deep in water. We'd always laugh our asses off. Karl would call us a "couple of skunks" and tell us to get back to work.

By then there were quite a few of us in the younger generation. My cousins Jan Eiven and Stan were working with Uncle Karl. The Nes brothers, Davin and Jeff, were sons of Magne Nes, a highliner from Karmoy. Johan Mannes was the son of my dad's good friend Borge Mannes, and Kurt Jastad was the son of his friend Jan Jastad. The old guys spoke Norwegian, and we young guys answered in English. My brother Norman began working with us on the *Northwestern* and I believe, for a while, we had the youngest crew of any crab vessel in the fleet.

As I grew older, I didn't want to be a kid anymore. I wanted even more to join the world of men like my father. On the occasions where a few boats were docked together in port, the old guys gathered in the galley to smoke, drink, and tell stories. Back then they had eight-track tape decks, and these guys loved to listen to Johnny Cash and George Jones in constant rotation. At times, someone would break out a guitar or an accordion, and they'd howl old Norwegian songs and country-western tunes. Lloyd Johannessen remembered that kids weren't allowed to sit in the galley, so he'd sit outside. When the men needed a refill or ran out of mixer, they'd holler for Lloyd to run down to the lazarette and grab bottles. Sometimes someone would mix him a drink and sneak it out the window. Still, I felt on the wrong side of some invisible line that distinguished men from boys.

When we were finally welcomed at the table, I felt as if I'd passed some rite of initiation into manhood. However, the hazing continued. I remember one time drinking a beer at a horseshoe table with five men on either side of me, and suddenly I realized I had to take a leak—and quickly. I was embarrassed to ask them to get up—I didn't want them to think I couldn't handle drinking beer. I said I had to get out, and of course they refused to get up. "Crawl under the table!" someone shouted, and they all began to laugh. Then I realized I would have to push back against the hazing, or it would never stop. So I stood up in my seat, climbed up on the table, marched across it, and jumped off and headed for the can. The men were amused by my cockiness and burst out laughing.

That same summer in Bristol Bay, an older guy named Sven kept teasing me. I decided not to take it anymore. One night when he was asleep I punctured the soles of his boots with a pick hook—a tool used for picking fish from the net. The holes were so tiny he couldn't see them. The next day he kept complaining. "My feet are soaked!" This went on for days. When John Jakobsen finally figured out what I'd done, he got a huge laugh out of it, but he never told Sven, who would have killed me.

When I was sixteen, another of my early attempts to become my own man backfired. My whole family was on vacation in Norway and when I finished salmon season I was supposed to get on the *Northwestern* for blue king crab. But I'd met a girl over in Karmoy the summer before, and I figured I could get a job over there and chase this girl around. I decided to surprise my family. I jumped on a plane for Norway, then showed up at my grandmother's house. My mom just about had a heart attack. Her jaw dropped. And my dad—he was not impressed.

"Ah, you dummy," he said with a smile. This was one of his favorite expressions, and one that he jokingly called his three sons when we disappointed him. He thought I should be in Alaska making money. I had signed on, and he thought I was being lazy. He didn't want to create a spoiled punk, acting flamboyant, like a big shot, and he was embarrassed. "If you say you're going to be there, then be there," he lectured. "It costs a lot of money to buy a ticket to Norway. You could be making good money!"

He didn't send me back to Alaska, though. He knew I was making a mistake, but he let me make it. That's the kind of father he was. He wanted me to learn lessons the hard way.

And I did. When I got back to Seattle, Fritjoff told me he'd made twenty grand on the *Northwestern* in just a few weeks. With my half share I could have made ten grand. He rubbed it in. As it turned out, I didn't even get the girl, either. Lesson learned.

At that time, I believed the single thing that prevented me from becoming an adult was being forced to go to high school. It meant I'd have to miss the best crab fishing months of the year—and for what? I wasn't motivated in school, and my parents never pushed me to go to college. To this day I hardly know the difference between a master's degree and a bachelor's. So I resented having to be in high school.

During these summers, I would return to Seattle with more money than the teachers made. A football coach tried to get me to try out for the team. I looked at him and said, "Would you rather play a game, or would you rather spend your summer making money?" I'll never forget the look on that guy's face. He was speechless. It was odd to realize that I was the one telling an adult to get his priorities straight. One teacher actually asked me how to get a job on a fishing boat. I remember thinking, You're *asking* me *how to get a job in Alaska? Shut up and teach!*

As far as I could see, the only worthwhile thing about high school was auto shop, where I could work on the cars I'd bought with my summer earnings. My first car was an El Camino I'd bought for five hundred bucks and fixed up. Once I started making cash I bought a Mustang Mach One. Norman bought a Camaro. We'd drag race those things up and down the street and tear the town apart. My car had racing slicks and open headers. Once the local magistrate came to recognize my car, I had to paint it from green to blue so I wouldn't draw as much attention. Cars were definitely more important than school. Then when I saw Fritjoff drive up in a brand-new Corvette, that sealed it. I wanted out of high school.

In our family, nobody had a high school diploma. My dad only had seven years of schooling and he really wanted me to graduate, so he offered me a deal: If I finished school, he'd give me a full-time job on the boat. If he didn't have room on the boat, he would find me a job on another boat. It was so strange: All I wanted was to be like my father—who'd never been to high school—and here he was asking me to complete it. If I had said no, he wouldn't have forced me, but he thought it was the right thing to do. It meant a lot to him, more than it did to me. So I stuck with it.

The Hansens are not the first Norwegians to set sail in search of adventure and riches. It's in our blood, and goes back hundreds of years. Starting about 800 A.D., the northern seas were ruled by fierce warriors called, of course, the Vikings. The image of the Viking with his long sword and horned helmet is as embedded in world culture as cowboys and Indians, Trojans and Spartans. When you think of the pro football team, or Viking stoves, or NASA's *Viking* spacecraft that landed on Mars, you rarely think about the Vikings, the people these things were named after. Sorry to break the news, but they probably didn't actually wear horns on their helmets. Even so, they were some tough characters. As soon as they reached manhood, the men would set out in longships and go raiding other villages. They had huge feasts, slaughtered bulls and sheep to eat, and drank ale and mead wines from rams' horns until they were sick.

The word "berserk" comes from the Vikings. The berserkers were warriors who wore pelts of bears or wolves, worked themselves into a crazed, trancelike fury, and were then unleashed by their kings to hack their enemies to pieces with swords, spears, and battle-axes. You didn't want to mess with the berserkers.

The Vikings were great naval architects. Their longships were narrow, graceful, and built from the lightest wood possible. They had a shallow hull, so they were very fast, and could land directly on beaches if necessary. When winds were favorable, the longship sailed with a big square sail. When winds were unfavorable, the crew rowed. Depending on the length of the ship, there were anywhere from twelve to sixty oars that stuck out from holes below the gunwales on either side. One ship, the *Long Serpent*, had thirty-four oars on each side, so it took sixty-eight men to row it. The boats were symmetrical—the bow had the same dimensions as the stern—so they could be rowed in either direction. If the seas at their stern became too big, they could instantly switch and buck into them, without risking a capsize by coming around and lying sideways in the troughs between waves.

Some ships had a dragon head carved from wood mounted on the bow, and the dragon's tail mounted on the stern. Since it was bad luck to face a dragon toward your home port, you were supposed to re-move the head as you approached. The ships were so valuable that when one band of Vikings attacked another, usually the first prize the victors took was the losers' boat.

Whether we know it or not, when we talk about boats, we're pay-ing debt to Viking innovations. For example, the word "starboard"—the right side of a vessel—comes from the Old English word *stéorbord*. *Stéor,* which means rudder or steering paddle, and *bord,* which in those days meant the side of a ship. The Swedish/Danish equivalent was *styrbord,* and the Icelandic was *stjórnborði*. Back then the tiller was always on the right, so that a right-handed man could face forward and steer with his right hand behind him.

The most famous Vikings are Erik the Red and his son, Leif Erikson. Erik, who was named after his red hair and beard, was born in Norway around the year 950, in Rogaland, the same county as Karmoy. Erik had a hot temper, and at an early age he and his father were thrown out of

Norway for murder. They sailed their wooden ship to Iceland, almost 600 miles across the Norwegian Sea, and settled on the west side of the island. However, Erik just couldn't control himself. After he killed a few more people he was also exiled from Iceland. So he kept sailing, another 150 miles to the northwest, this time reaching an unknown land that now is cold, barren, frozen with glaciers, and lacking in timber and game. Back then, scientists now believe, it was much warmer and a more hospitable place. Still, it was a tough place to settle, but it was all he had. Erik figured that if he gave the place a decent name, more people would join him. He called it Greenland—the first European colony in what centuries later would be called North America.

The settlement thrived. Iceland was overpopulated and Greenland offered a new frontier. As many as four thousand people lived there. Erik named himself chieftain and became rich, powerful, and respected. Around that time Christianity was starting to spread across Scandinavia. Erik was a pagan, but his wife converted and built a church in Greenland. Still he was lukewarm about Christianity. After her conversion, the sagas say that his wife refused to sleep with him, "much to his displeasure." No wonder he stayed a pagan.

Leif had the same adventurous spirit. He was born in 970 in Iceland, before his old man was exiled. So when he was a boy he packed up and moved to Greenland with his father. Like his dad, Leif was a great sailor. On one voyage he came across a shipwreck, with the crew still alive, clinging desperately to the swamped boat. He rescued them all, and after that, the people called him Leif the Lucky.

In Greenland, there were legends of more lands to the west. In the year 1002 Leif Erikson set out to explore them. The sagas say that Leif tried to convince his father to join him. Erik was getting feeble and couldn't stand the wet and cold like he used to. He didn't want to go, but finally Leif talked him into it, so Erik packed his seabag. As he rode down to the boat, his horse stumbled and threw him, and he injured his foot. Erik the Red took it as an omen.

"I am not intended to find any other land than this one where we live now," he told his son. "This will be the end of our traveling together."

Leif set sail without his father. He and his men sighted some inhospitable shores—probably what we now call Baffin Island and Labrador—before landing on a warm, sandy beach surrounded by ample forest and game. They made camp and sent out scouts. One of the men returned with a thrilling discovery: wild grapes, an exotic fruit that the men had heard about but never actually tasted. They stayed the winter, finding that the ground didn't freeze; they ate the wild grapes and the abundant salmon. The grass stayed green all winter, which could sustain their livestock. There were maple trees and wild wheat and dark-skinned native people in canoes. Leif called the place Vinland.

After the winter, Leif returned home and told of his discovery. By then his father had died in an epidemic. Leif's settlement at Vinland didn't weather as well as his father's in Greenland. Soon after his return, Leif's sister Freydís Eiríksdóttir sailed to Vinland on an expedition led by seasoned sailors. Leif gave her permission to live in the buildings he had left behind. Once they arrived, Freydís proved a tough customer. During a skirmish with the natives, she berated her male companions. "Why do you flee such miserable opponents?" she cried. "Had I a weapon I'm sure I would fight better than any of you." It didn't matter that she was pregnant at the time. To prove her point, she snatched up a sword from the body of a slain Viking and rushed at the natives. Then she opened her shirt, exposed her breast, and slapped it with a sword. The sight of a pregnant she-Viking slapping her bosom with a broadsword struck fear into the hearts of the Indians. They fled to their boats and paddled away.

Freydís was a vengeful woman. The sagas say that she argued with the men leading the expedition. The disagreement raged, and finally she ordered her crew to murder the others. They did the deed. All the opposing men were executed, but the five women were spared. Freydís's soldiers refused to kill them.

"Hand me an axe," she said.

Freydís murdered the five women. The next spring she sailed back to Greenland. "I will have anyone who tells of these events killed," she told her men. "We will say that the others remained behind when we took our leave." On their return home, however, word leaked to Leif. He tortured some of his sister's men until they confessed all the details.

"I am not the one to deal my sister, Freydís, the punishment she deserves," he said. "But I predict that her descendents will not get on well in this world."

Leif Erikson died around 1020 of unknown causes. Historians may never agree on the exact location of his Vinland settlement, but it was probably in Newfoundland where archeologists have found a Viking village. From that camp, the sagas say that the Vikings ventured as far south as Cape Cod and Long Island. In other words, the Norsemen explored and settled the North American mainland almost five hundred years before Christopher Columbus arrived. That's why Norwegian sailors are so proud of their heritage. Ask any Norwegian and they'll tell you: We're the ones who discovered America.

In a family of all brothers, there was a lot of competition. We all wanted to prove to our father that we were good enough. I guess sibling rivalry is in our genes. It was in Dad's and Uncle Karl's. As adults they lived just a few blocks apart, but they'd go months without talking. When they did get together during the holidays or at a party, they

Norman Hansen on an icy opilio fishing trip on the *Northwestern,* circa 1982. *(Courtesy of the Hansen Family)*

would argue about something that happened twenty years ago, or who caught the most crab. To an outsider it might look like they didn't get along, but actually, they had the incredibly close bond of brothers. They did argue, but at the end of the day they were the best of friends.

My brothers and I inherited that. The main difference is that while Dad and Karl had their own boats, Norman and Edgar and I are bound together on the *Northwestern,* cramped in close quarters for half the year. Edgar and I still butt heads all the time. If you've seen any episode of *Deadliest Catch,* you know that.

A lot of people don't even realize that Norman is onboard with us. While Edgar and I sometimes ham it up for the camera, Norman will have none of it. He's a classic Norwegian fishermen, shy like some of Dad's friends from the Old Country, and won't say a word until someone drags it out of him. When a camera is pointed at Norm he just glares back, silent. Usually the cameraman gets the point and gives up. If he persists, Norm tells him, "You're wasting your tape." If they still don't get it, Norm just turns and walks off. In five years of being on television, Norman has logged about ten seconds of screen time. I can respect that. He is his own man.

Growing up, Norm and I were just a year apart in age, and yes, there was rivalry, probably because we were very different. Norman was extremely smart—ingenious. He would latch onto some hobby—electronics, coin collections, gun collections, locksmithing—and totally master it. He would get intense about one thing, and then another. His mind was always working. He was one of those kids who would make a vodka still in his closet. Once he blew the transmission out of his car when he was racing around the neighborhood, and had to drive it in reverse to get back home. The next day was prom and Norm had a date and needed the car. So he recruited our friend Mark Peterson to help him swap in a new tranny. Norm didn't know how to do it—he didn't even have a tranny jack—but that didn't stop him. He could figure anything out. So Peterson was under the car, squirming under the weight of this hunk of steel. Norman stood poised with his wrenches and said, "Just push it up and hold it," as matter-of-factly as if he'd asked you to refill his coffee. Peterson thought the thing would crush him, but Norm was calm and cool. That's how he is. Whatever he sets out to do, he does it.

Edgar, being the baby, got a bit more affection from Mom and Dad. They were more huggy with him. Since he was four years younger than Norman, the competition between the two wasn't as intense. Maybe by then, Mom was just worn out. Edgar would say he was the black sheep of the family—and it's no wonder. We used to tease him that he was adopted because he had brown hair and the rest of us were blond. He took after our grandmother Emelia.

Typical of most schools, ours had the big cliques: jocks, rockers, stoners, and preps. Edgar was in the smoking section—he was a rocker. He hung out with the wrong crowd, had long hair and a guitar, and had a fast car, a '69 Mustang Mach One Fastback—the one I gave him. Edgar skipped so much school that he had to go to an alternative high school. According to him, "It was like a day care for adolescents."

As for me, I became the evil older brother. One year in elementary school Mom gave us lawn darts for Christmas, the long ones with the steel tips. Big mistake. I had a heyday with them, turning my brothers into moving targets. We got boxing gloves and had competitions, but since I was the oldest and biggest it wasn't always fair. I was just an ass, so we had our clashes.

As Norm and I got older, we had our battles. One time in Dutch Harbor I managed to overdo it in town and passed out on the galley floor. Norman thought it would be funny to pour some beer on my face, revenge for the time I had done the same to him. When the beer hit my face I snapped out of it and went after him, throwing punches. The fists flew from the galley, through the entryway, through the door, onto the deck through the snow and mud, and all the way to the stern. We kept pounding on each other until we were too tired to keep going. Every time he slugged me he shouted, "I hate you." All the men on the boat just watched, and laughed, and let it happen. That's how it works up here, and it was probably a good thing—I'm sure I had it coming. That was the last time Norman and I fought.

Edgar and I had some conflicts, too. When we decided to write this family history, Edgar told his version. Here's him talking, "I couldn't stand Sig. Usually you think your older brother would stand up for you. But no! I was getting bullied at school. I used to wear these big brown glasses, and was getting teased on the way home from school."

He came to me and told me the bullies were throwing rocks at him.

"Would you beat them up?" Edgar asked.

According to Edgar, I just started laughing. "Take care of it yourself," I said. This was the same advice I'd learned in the Sverre School of Hard Knocks. My father used to tell me when I got a black eye, "If you can't win, get a bigger hammer, and hit the other guy first."

The torment continued for years. Finally Edgar got sick of it. Edgar remembers, "I was walking down the stairs at home, and Sig was sitting on the fourth step from the bottom. I think he was on the phone. And as I passed he grabbed my leg. He wouldn't let go. Started laughing. I got pissed off. And his head was right behind my elbow, so I gave him a crack. Knocked him in the teeth. He went *ballistic*. He threw me on the ground, ran into my bedroom, and slammed the door shut. He must have been in there twenty minutes, ripping my room apart. Literally tearing down the shelves, breaking the bed. Everything. Even now, if you look close, you can see he has a little V right on his front bottom teeth, if you can see past the smoke stains. He always has a reminder of the time his little brother fought back."

Looking back I feel sorry for my mom. It was just us three brothers, and we were all a bit rough around the edges and independent. We can thank Dad for that—a free spirit and his own man. He was often at sea, and when he was away, Mom had to do the work of raising us by herself. On the outside she appears prim and proper, but I realize now what a strong lady she was. She had to learn to fend for herself. Like most fishermen's wives, she had to worry about her husband coming back safely. She had to live in a foreign country and learn a new language. She had to manage the house. She had to make sure the kids got to school. She also made sure we went to Sunday school, and later joined the youth group at Rock of Ages. We were all believers, but like most teenagers, we had other interests, and rebelled. One day the group came over to our house for a social event, and ended up packed into Norman's bedroom listening to Iron Maiden's *666*. The counselor was pretty disappointed. What the hell could he expect from kids like us?

Then everyone was leaving to fish. All three of us brothers followed in Dad's footsteps to Alaska. My mom was strong and understood it was required for the job, but still it had to have been hard. I never looked at things from her perspective. Even when the family dog died and she was upset on the phone while I was working on the boat in Alaska, it was hard to be sensitive. I was thinking, *It's just a dog. What's the big deal?* Then she started to cry. It may have been the first time I'd ever heard her cry. I didn't now how to respond. I felt like I needed to be strong for her, so I showed no emotion. I also worried that if I was too sympathetic I'd get homesick, and that's the worst thing when you're fishing. Now that I'm older I've got a lot more empathy, and looking back I see I could have done a lot better. I should have called home more, but I didn't. I stay in touch with my wife a thousand times more than I did with Mom back in the day.

We were a different kind of family. We weren't like the modern American family where everyone shares their feelings and tells each other how much they love them. Dad rarely gave us a hug. That's just not who he was. We weren't a cold family, though. We knew that our parents loved us—it just wasn't expressed all the time. Much later in life, I remember asking my dad about it. "Well, nobody hugged me when I was a kid," he said. I think he wanted to, he just didn't know how. I don't hold it against him.

When I was younger, the whole idea of saying you loved someone was foreign to me. When I was eighteen I was fishing out of Akutan, an island forty miles east of Dutch Harbor. My friend Johan Mannes and I wanted to call home. We had grown up together, and in the off-season shared an apartment in Seattle. There were only two pay phones in the whole small town and we had to wait outside where it was snowing. The glass booths were broken. We froze as we waited to use the phone, but we were leaving in the morning and it was our last chance to talk to anyone for a couple of weeks. Finally Johan got on and did his thing, telling his mom that everything was fine and he'd be home soon. He ended the call by saying "I love you, Mom."

"What the hell did you just say?" I asked. "Are you nuts?"

Finally he got sick of me. "What's the big deal?" Johan said. "You should say that more often, asshole."

I ignored him and called my mom. "Everything's fine," I told her. "Fishing's good."

The next thing I know, Johan shoved his way into the phone booth. He punched me in the ribs.

"Say it!" he said under his breath, so she couldn't hear it. Wham! He punched me again, harder. "Say it!" *Wham!* "Say it!"

So I said, as quickly as I could, almost in a whisper: "Okay, love you, Mom."

I could hear her choke. "*Ja, ja,*" she said, nervously, in her Norwegian accent. She'd never heard that from me. "Love you, too."

Click.

Norway is larger than the state of Washington, but shaped like a spoon, more than eight hundred miles long and just a few miles wide in parts of the mountainous "handle" that extends north to the Arctic Circle. It has about eight thousand miles of coastline and thousands of islands. Much of the interior is uninhabitable, so the early settlements were on the coast. Nowadays you can drive across the country in tunnels that burrow through mountains, but long ago the only way to get around was to sail. With the navigation techniques of the Middle Ages, the Vikings could sail the North Sea east and west safely, even in dense fog. Yet they needed landmarks to navigate north and south. So they hugged the western shore, a route that funneled them into the narrow Karmsund Strait between the mainland and the island of Karmoy. The king saw that with all the boats concentrated in the strait, he could charge a toll. He put a duty on all the hunting game going south, and all the corn and wheat going north. They came to call this passage the "road north," which translates as "Nor-Way."

From Karmoy it's closer to the Shetland Islands in Scotland than it is to Oslo, the capital of Norway. According to legend, the first village was founded when two sisters from Scotland were lost at sea and drifted east across the North Sea. Once they landed, they married local fishermen, from whom all residents descend. The island has been inhabited for four thousand years.

Karmoy's sailing lore dates back to before the Viking era. According to Norse mythology Thor, the god of thunder, waded across the Kormt River each morning on his way to Yggdrasil, a massive ash tree that was known as the tree of life, the center of the universe. Local legend says that the Kormt River was actually the Karmsund Strait.

Karmoy was the home of King Harald Fairhair, the first to unite all of Norway under one king. Harald was a badass. Before they called him Fairhair, they called him Tanglehair, because he made an oath not to comb or cut his hair until he became sole king of Norway. He was king for seventy years. When he got old he appointed his sons rulers of his kingdom, and assigned the oldest, Eirik the Bloodaxe, as king over them all. However, that didn't work out. As soon as Harald died and was buried across the strait, his sons began arguing over who was in charge. This led to a battle the very next year at Tunsberg, where Eirik killed his own two brothers, Olaf and Sigurd. Tough break for Olaf and Sigurd.

Yet Eirik wasn't out of the woods. Another brother, Haakon, who had been living in England, arrived with a great army. Eirik saw the writing on the wall. He fled to England for the rest of his life, while Haakon became king of Norway.

The age of the Vikings ended when Christianity arrived. There is an old stone church in Karmoy called St. Olav's that dates back eight hundred years. On the side of the church in the grassy cemetery, a twenty-foot spire called the Virgin Mary's Sewing Needle leans toward the church; but it doesn't touch. The sagas say that once the needle hits the wall, doomsday is here. Over the centuries, gravity has forced the spire ever closer to the wall, so the monks have kept chipping away at the wall. Still, the two have yet to touch.

The Karmoy coast is dotted with seafaring villages like Akrehamn, where my family comes from. Fishermen have been setting nets for herring for centuries, largely untouched by the advances of the modern world—there wasn't even a bridge to the mainland until 1955, when my dad was seventeen years old. My great-grandfather Johan Hansen was a herring boss, or what they called a "fishmaster." During hard times in the early 1900s, he helped the widow of a cousin by buying part of her farm for three-hundred crowns—about fifty dollars. His family lived on that seven-acre parcel for many years.

My grandfather, Sigurd, was born in Akrehamn in 1914. He began fishing as a boy on a seine netter with his father. A seine net is a big net that hangs vertically in the water. As the net is released from the ship, a wooden skiff drags it in a circle around the ship. Then they tighten the bottom of the net, like pulling the drawstring on a purse, to trap the fish inside. It's sometimes called purse seining. My grandfather was sixteen when he started rowing the skiff while his dad sailed the seiner. Of course both boats were wooden, and neither had a motor. They worked from six in the morning till ten at night, rowing that skiff around the entire island, about a forty-mile trip.

My grandfather *(first on the right)*, Sigurd, as a crewman aboard the *Holma* in Norway in 1935. *(Courtesy of the Hansen Family)*

They fished herring and cod for many years. Later, Sigurd was on a ship called the *Vigra,* and the fishing was so good that they accidentally sank it. What happened was, the herring were getting soft, ready to spawn. The fishermen kept filling the tanks. In the holds, planks ran lengthwise to keep the fish from sloshing around. These planks were supposed to be held in place by an H-beam that sat on top of them. Apparently someone forgot to bolt the H-beam. The herring didn't settle like they were supposed to and knocked the planks loose. The fish liquefied like soup, rolling from port to starboard, starboard to port, until finally the boat laid over. Sigurd and the crew hauled some oil pumps and other heavy equipment to the other side of the boat. It worked—the boat righted itself! Yet a moment later it flopped down the other direction. They kept hauling the gear from side to side as the boat rolled to and fro, but finally it swamped. It was hopeless.

Sigurd and the crew abandoned ship and got on the skiffs. They sat there and watched the *Vigra* slowly submerge, only its pilot house above water. As she sank, a board or something must have got caught in the horn, because she let out a long whistle, like she was saying good-bye. Then she went down.

My grandfather used to tell us other stories, too. Another time Sigurd was on a boat called *Kval*—Norwegian for whale—when it caught fire. They put out the fire and got towed to safety, but lost the boat anyway. After that, when they told the story over and over, they called it "the burning whale." On another boat a man died onboard, but the fishing was so good that rather than take him in to port, they put him on ice until they finished the trip.

During the Great Depression these men went to Iceland to fish. Sometimes they would make three trips a summer. The weather was cold, foggy, and miserable. When the market collapsed, nobody wanted to buy any herring. They worked their tails off and didn't get a dime. Years later, when my grandfather was an old man—after he'd fished thirty-four seasons in Iceland—he would still joke about it. "I *still* haven't had my settlement!" he used to say.

My last year of high school, something happened that changed the course of my life—though I didn't even know it at the time. That summer I wanted to spread my wings and get out. I told the Old Man I didn't want to work for him anymore. I wanted to work for Oddvar on the *Silver Wave*. It was a highliner that made a lot of money. Oddvar hired me for blue crab season and everything was cool, but the day before I was to fly to Alaska, I called Oddvar to ask what day he needed me.

"I don't think I can take you on this time," he said.

"What?"

He hemmed and hawed, said he had a full crew, and apologized.

By then it was too late—I already had my plane ticket. I showed up in Dutch Harbor without a job. I had already given up my spot on the *Northwestern* and I was too proud to ask for it back. So I went over to a boat called the *Pacesetter* and got hired as the cook. Just before we launched, I jumped over the railing and sprained my ankle. I had to stay in Dutch. I guess I was lucky. The *Pacesetter* sank after that. No survivors.

Then I had to make the humiliating call to my Old Man.

"Boat full?" I said.

"*Ja*," he said in his thick accent.

"There's no spot with the *Silver Wave*?"

"We'll get you on here."

I had to go back to the Old Man's boat. I've worked on the *Northwestern* ever since, more than thirty years. I never could figure out why Oddvar didn't hire me, though.

Damn near years twenty later, after Dad had passed away, I ran into Oddvar and he told me the real story. Turns out that he and the Old Man had been talking in Ballard the summer I was supposed to work on the *Silver Wave*. The Old Man put the screws to me. He told Oddvar not to hire me. He already knew where he wanted me, but he didn't want me to know. He kept it all secret. Had I gone over to the *Silver Wave,* I might never have gone back to the *Northwestern.* The Old Man didn't tell anyone why he wanted me on his boat. Maybe he wanted me to be skipper all along and was just setting everything in motion. I guess you could call it my destiny. By the time I knew enough to ask him—and thank him—he was already gone.

2

KARMOY DREAMS

My father's ship, the *Foremost*, was not what you'd call a state-of-
the-art vessel. Built in 1945, the eighty-foot wooden sardine
boat was designed for trawling and seine-netting in gentler waters, not
for hauling steel crab pots in the Bering Sea. Its timbers, soaked in decades of oil and diesel, could combust in a split second. What's more,
the boat leaked like a sieve. Each spring they'd pull it up in dry dock
and fill the holes in the deck and the hull with cork, but the repairs
never lasted. Anytime I start to complain about the *Northwestern* being
thirty years old and in constant need of repairs, I just remember the
sort of vessels Dad dealt with, and I count my blessings.

As captain of the *Foremost*, Sverre was not going to complain. He
had moved from cabin boy to cook to full-share deckhand, and now
for the first time he was skipper. He'd taken the job when he was
twenty-nine, one of the youngest skippers in the fleet, and two years
later he was turning a tidy profit. The old sardine boat was owned by
the Wakefield Seafood Company, which in those days had a stake in

The wooden *Foremost* as it appeared in 1945, before being converted for crab fishing. *(Courtesy of the Hansen Family)*

a lot of the Bering Sea operations. Although it was an American vessel owned by an American company, the *Foremost* had seen a run of Norwegian skippers, many from Karmoy. Its crew roster reads like a Who's Who in the history of crab fishing. Before Sverre, the captains had been pioneers like Sam Hjelle, Magne Nes, and John Johannessen. Other future highliners like John Sjong, John Jakobsen, and my Uncle Karl worked on deck at one time or another. These were the guys we all idolized as kids. "A great boat," John Sjong says today. "I loved that boat."

Sverre was ambitious, and being captain wasn't enough. He wanted to own a boat. To do that, he'd have to make a lot more money—and catch a lot more crab—than he'd done so far this 1969 season.

The previous spring Sverre and Wakefield had replaced the old Atlas engine with a new high-speed Caterpillar diesel. The new engine ran much hotter, but the men didn't insulate the exhaust stack. Sverre's best crewman, a Karmoy man named Sigmund Andreasson, deemed it an accident waiting to happen.

"I refused to go," said Andreasson. "I didn't trust the boat. Sverre was pissed that I wouldn't go, because it meant he had to find new crew."

Sverre replaced Sigmund with another Karmoy man. Magne Berg was a big, strong, husky fisherman who'd immigrated to New Bedford, Massachusetts, as a kid with his parents. He grew up on the scallop boats. He was a hell of a nice guy. Fearless. He'd welcome a brawl, and he wouldn't back down. One time a crab pot fell right on top of him. He lifted off the whole damn thing by himself. "Who the hell dropped that pot?" he yelled, and kept on working. And damn if he wasn't quick. He picked gear so fast that they'd pass other boats standing still, and the skippers would call on the radio, "Are you running or are you fishing?"

Captain Sverre and Magne Berg and two other hands had worked the *Foremost* all season, since steaming across the Gulf of Alaska in May. It had been a long season. They couldn't find the crab. They weren't making a lot of money. Same old thing everyday—pulling pots with just a handful of crab. Fishermen were restless and pissed off. Then came the strike.

There had been a strike two years earlier, but it had backfired. After all that effort, they got paid only one more penny per pound. Even so, the powers that be thought they could get it right this time. They called another strike. Word spread across the fleet, and the fishermen had no choice but to obey. Some of the owners who had a lot at stake—bank loans and boat mortgages—wanted to hold out for a better rate, but the deckhands and the nonowner captains just wanted to fish. They hated the strike. They just sat on the boats all day with nothing to do. They got itchy feet. If they had a few dollars they would go up to the Elbow Room and have a few. If they could afford the fare they'd call it quits and fly home to Seattle or Norway.

Finally the strike was settled and the fishing resumed. By December the ranks were thin. Two of Sverre's crew left, leaving just him and Magne. Some of the skippers were going home, too, leaving their crews behind. One morning, two Norwegians were walking the dock after their boss had split. Turned out they were Karmoyverians, Leif Hagen and Kristoffer Leknes. Sverre had known them for years, and besides, they were from Karmoy—they *must* be good fishermen. He hired them at full shares.

Leif Hagen was a year younger than Sverre, and they'd known each other in school. He was friendly, soft-spoken, liked to tell stories and laugh, and was one hell of a hard worker. He had come up in May on the *Admiral*. When one of the crewmen left for a better paying boat, Leif flew to Seattle and recruited his buddy Kristoffer.

Krist Leknes, as the Karmoy boys say, was all right when he was drinking, but an asshole when he was sober. He was outspoken. He was argumentative. He liked to play poker. He complained a lot. He'd ask what you wanted for breakfast, and if you said eggs he'd give you pancakes. "This is a fuckin' *boat*," he'd say. "You'll eat whatever I give you."

It was a good crew, as reliable as any. These four guys were the hold-outs, the ones who wouldn't go home yet. They wanted a bit more money. To leave now would be to admit defeat. They just needed one jackpot to end the year. Aside from my father, they were all bachelors and cared more about the extra cash than being home for Christmas.

It was a reunion of sorts for Leif Hagen and Magne Berg. Two years before, they had been working on the *Emerald Sea* when they got a taste of the dangers of Alaska fishing. They had just left Kodiak, where they stocked their sea store with whiskey, beer, and cigarettes, and motored a few hours toward Cape Barnabas. Leif was on watch when Magne climbed up to the wheelhouse and sat down.

"Well, aren't you going to relieve me?" Leif said.

"Yeah, just a minute. I'm going down and get me a bottle of beer."

"Okay."

When Magne returned he tossed a bottle to Leif. Then he opened his own.

"Screw it," Magne said. "We're sinking, anyhow."

What the hell? thought Leif. He checked the controls. The boat was running full bore. Everything felt fine. He figured his friend was just joking. Magne liked to mess with you sometimes. They sat there and bullshitted for a while, and finally Magne took the wheel.

"Bring me another bottle of beer before you go lie down," he told Leif. "And take a look in the engine room and see what's going on."

Leif climbed down the stairs of the cabin and then went below deck. *Jesus Christ!* The water was halfway up over the engine. He could see men in T-shirts diving into the water. When they came up for air, he saw it was the engineer and the skipper. There was a huge leak in the hull and—sure enough—the damn boat was sinking!

The men rushed to the deck and pulled the cord on the life raft, and it sprang to buoyancy. It was just a simple rubber donut with a floor, but no tent. The men tied it to the rail and it bobbed in the sea. Though the engine room was swamped, the galley and wheelhouse still held enough air to keep the *Emerald Sea* afloat. Those last hatches, though, could burst at any minute, and the boat would reach the ghastly tipping point where it weighed more than water. Leif and Magne leapt onto the raft and waited for the others. Magne palmed a knife, ready to cut the line should the boat go down, otherwise it would drag the raft down with it. They wouldn't even have time to untie. Suddenly the engineer ran out of the cabin.

"Hey, wait a minute!" he called. "Wait a minute!"

"Wait for what?" Leif called back.

"Wait till I get the sea store!"

Within seconds the engineer came running out with a huge seabag, six feet long and three feet across. It bulged with two cases of whiskey bottles and fifty cartons of cigarettes. *Booze,* Leif thought. *Jesus Christ!* The *Emerald Sea* was wholly swamped, and now Leif found himself climbing *out* of the life raft and *onto* the sinking ship—just to save the booze! Although the raft was tethered to the starboard, there was enough slack in the line to drift a few feet. He planted one foot on the rubber tube of the raft and stretched the other on the deck of the ship, straddling the cold ocean below him. The engineer heaved the sea-bag into Leif's arms—damn, it was heavy—and Leif humped it into the raft. Just as Magne pulled the bag safely onboard, a wave washed through, pulling the raft away from the ship's rail. Leif did a split, desperately trying to pull the raft toward the ship with his toe. He failed. His legs stretched as far as they'd go, and while his crewmates looked on helplessly, his feet gave way and he splashed into the icy water. Fully dressed in boots and raingear, he kicked and flailed, his heart pounding as his body instantly went numb. With all his strength Leif reached for the life raft. Magne was waiting with a strong hand. He hauled Leif into the raft, where Leif collapsed on the rubber floor, breathless and shivering. He realized that—*stupid, stupid, stupid*—he had almost traded his life for whiskey and smokes.

The engineer climbed into the raft, and now with the ship barely afloat, they waited for the captain. Captain's honor dictated that he either go down with the ship, or at least be the last one to abandon it. He had given up trying to stop the leak in the engine room and now was up in the wheelhouse calling mayday on the radio. How long would the men wait for him? They didn't want to untie too soon. As long as it was afloat, the *Emerald Sea* was the safest place to be. They knew stories of crews that had lost their lives in rafts while the ship they abandoned never actually sank.

The distress calls were eventually answered. Two boats arrived. Leif, Magne, and the engineer were pulled to safety. With the captain still onboard, the rescue boats attached lines to the *Emerald Sea* and towed her toward shallow water. They figured that even if she sank, the boat would be easier to recover there. Now it was a race against time. They needed to reach the shelf before the ship went under. Swamped and flooded, the *Emerald Sea* rode like a torpedo, just barely above water. As they reached a steep slope that approached shallow grounds, the captain abandoned ship and was rescued by the others. They chopped the lines and watched the ship sink, a nauseating boil of bubbles rolling over the water. Instead of resting in the shallows where it landed, the boat slid down the slope to the icy depths. The *Emerald Sea* was never recovered.

Maybe after one lucky rescue, Leif and Magne should have called it quits—or at least seen themselves as jinxed and agreed to not work together anymore. But that wasn't how Karmoy fishermen thought. They didn't get philosophical about life and death. It was just part of the job.

Two years later Leif and Magne were reunited on the *Foremost*, and on a cold December morning they steamed four hours northeast from Dutch Harbor over the North Head of Akutan Island to Akutan Bay, where Sverre hoped his luck would be better. As Sverre watched the wheel, Cap Thomsen came over the radio from Dutch Harbor. Niels Thomsen was an old Dane who had wrapped fish at Pike Place Market as a young immigrant boy, sailed around the world, and sunk a Japanese submarine from a Coast Guard ship during the war. He was now owner of the Aleutian King Crab Company, and was always cooking up some gimmick for business.

"Santa Claus is offering a big price raise for Christmas," Thomsen broadcasted. "Come sell to me!"

But Sverre didn't have the luxury of holding out for a higher price. Since the *Foremost* was owned by Wakefield, he was obligated to deliver the crab to Wakefield's processing boat, the big steel *Deep Sea*, anchored in Akutan Harbor. Akutan was a tiny village on a tiny island, an active volcano. There was little more than a store and a bar—about forty structures and seventy-five year-round residents.

In those days, they didn't have the sodium lights that we have today, so fishermen couldn't fish at night. They would bust their ass from dawn until dusk, but then find shelter in a harbor and sleep. They also didn't have the hydraulic rack that we use nowadays to launch pots. They had to push the pots overboard by hand. In addition, they didn't have the hydraulic coiler. One guy had to coil by hand—a back-breaking and exhausting task that ripped up their hands and wore out their arms.

They couldn't find a damn crab. They went from one place to another, pulling their pots. Blanks. Empty. Finally, Sverre ordered the men to stack a few of the seven-by-seven pots on deck. On an old wooden sardine boat, you couldn't hold a lot on deck, but he loaded a half dozen and dropped them by the beach in a small cove.

After a couple of hours they pulled a test pot. It was already half full, the big red bugs undulating in a mass. *Boy oh boy,* Leif thought. *Here we're going to get a Christmas present.* Then, just when it looked like they were finally going to get a break, the throttle went unresponsive in Sverre's hand. He rushed down to the engine room and deduced that the accelerator was broken. He and the crew could jerry-rig something to get back to port, but this repair required new parts. They were forced to leave the bulging pots in the water and limp back to Dutch Harbor.

The part they needed had to be flown in. So they sat on the boat, while the wind blew and the snow fell. It was freezing cold. It was going to take a couple of days. Leif got so bored that he decided to clean out the forepeak, the cramped storage area at the front of the cabin. It was filled with a huge pile of old junk: oilskins, nets, and boots, all smelly and mildewed.

"Is there anything here you want to keep?" Leif asked the captain.

"Throw the shit away," said Sverre.

Leif dove in and started hauling junk out of the cabin and heaving it overboard into a pile on the dock. Then he came across a big suitcase. *What the heck?* Leif hauled it out on deck and popped it open—it was the life raft! Well, that was a major screwup, having the thing buried where no one could find it. He called over to the others and they all had a good laugh.

"A lot of good it would have done us buried in the forepeak!" Sverre cried.

Instead of throwing it away, Leif returned the raft to its proper place, atop a wooden box on deck behind the cabin where it would be easy to reach in an emergency.

Leif warned the others, "If any one of you puts so much as a rock or a piece of line on top of that thing, I'm throwing you overboard."

On the third day, the mechanic fixed the accelerator. A quick celebration was in order.

"To hell with it," cried Sverre. "Let's go up to the Elbow Room for a drink!"

"Okay," said Magne. Leif and Krist didn't have any money, so they stayed behind. Sverre and Magne climbed off the dock and there on the road was a pickup truck. They knew it belonged to Carl Moses, who owned the hotel and hardware store in Unalaska. Sverre peeked inside and saw the key in the ignition. *What the hell?* They jumped in and drove over to the Elbow Room. It turned out that Carl Moses was up in the window of the Alyeska cannery and saw the whole thing unfold. He called down to the Elbow Room on the CB radio, "I just reported a stolen car, and they better return it as soon as possible, because the cops are out looking for them."

Next thing you know, Sverre and Magne came flying down to the boat.

"Throw the lines!" cried Sverre. "Throw the lines!"

They jumped aboard, Leif and Krist threw the lines, and off they went.

Yet the *Foremost* would only last a day or so before the accelerator again failed. They returned to Dutch Harbor, aware that each gallon of wasted fuel would come out of their settlement; not that there would be a settlement as long as those crab stayed in the pots and not on the boat. They tied up at the oil dock and figured they'd fuel up in the morning when it opened. Sverre went to sleep in his stateroom. A few hours later, Leif heard someone boarding the ship. It was the Dutch Harbor policeman.

"Where's Sverre?" said the cop.

"He's sleeping," said Leif.

"Tell him to come down."

Leif climbed the stairs to the stateroom.

"Sverre," he called. "The cop wants you."

Sverre shook open his eyes and rubbed his temples.

"Get out of here!" he said finally.

"Do you want me to send him up?"

"No, no, no. I'm coming down." Of course he thought he was in trouble for stealing Carl Moses's truck, but he'd been caught and there was nothing to do but turn himself in. Sverre threw on some clothes and boots and stepped out on deck. He found the cop was in no hurry to arrest him.

"Do you want to do me a favor, Sverre?" the cop asked.

Sverre agreed.

A Russian freighter had pulled into Dutch Harbor earlier that day with a sailor who had a broken back. The cop's idea was to lower the patient onto the *Foremost*, take him into the oil dock, and from there they could drive him to the airstrip and get him to a doctor.

Sverre climbed to the wheelhouse and fired the engine and took the boat to the freighter. There was the cop, an interpreter, a priest, and a nurse. As they were lowering the patient off the freighter on a stretcher, a blond head poked through one of the portholes—it was a good-looking woman.

"How in the hell do you get any work done when you've got something like that on board?" called the cop. They all laughed.

They got the Russian aboard the *Foremost*, then delivered him to the dock. There they placed him on a flatbed truck and drove through the potholed roads to the airstrip. The drive must have done his spine more harm than any doctor could have done good. Anyway, the Russians were appreciative, and so was the cop.

"That's on me, Sverre," he said. "We'll just forget everything."

The next day they repaired the accelerator for a second time. Now the wind had picked up and was blowing 40 knots right in the harbor. Most of the boats decided to lie in port until it passed. The forecasts said it would calm down by late the next day, but Captain Sverre didn't have another night to wait. Besides, they didn't have that far to go. He figured they could motor to Akutan in the gale, and by tomorrow when it stopped, they'd already be at the gear.

They left Dutch Harbor in the evening of December 7, Pearl

Harbor Day. They steamed through the darkness. It was their third trip of the week from Dutch to Akutan. The tanks were filled with water for ballast in the rough seas, but there was not a single crab on board. They rounded North Head in the cold, howling wind and arrived at their gear between Akutan Island and the next island, Akun, in the middle of the night. They could see the shore. The gear was right there in Akutan Bay. They knew there were crab in the pots. It was just a matter of getting it on board before another mechanical failure.

Captain Sverre decided to get some sleep. He'd leave a man on watch to keep the boat jogging into the seas. In the morning, they'd wake up early and finally start pulling pots, make enough money to call it quits for the season, and head home to Seattle for Christmas.

Like I said at the beginning, the story of my brothers and me continues the saga of my dad and his brother. They paved the way for us in this country and in the fishing industry. That's why the story of Dad's fire aboard the *Foremost* is so important. All of the good luck that I've had, and any of the skills and judgment I have, were passed down by my family. It's almost like: They sank ships, so I don't have to.

In order to really feel what was at stake for my father as he set out for Akutan that December morning, you need to know who he was and where he came from. The hardship he was born into molded his entire life and molded the way he raised Norman, Edgar, and me. Although we were born in America, and never lived in Norway, we are Karmoy boys through and through. That's where our saga begins.

My dad was born in Karmoy in 1938, and his brother, my uncle Karl Johan Hansen, came along in 1942. The boys' roots on the island

The home in Karmoy where my dad and uncle grew up, with my grandmother sitting in front. *(Courtesy of the Hansen Family)*

went way back—their grandfather had been a member of parliament, one of the first to represent Karmoy. My father and uncle grew up in hard times. The Nazis had invaded and occupied Norway, and the Home Front government was exiled to England. German soldiers were housed in the local schoolhouse, and Sverre was old enough to notice them crawling around town.

Sverre and Karl lived on the small farm that their grandfather had bought. The property was way up the hillside from the harbor, a long walk for their father. There were no snow plows in Karmoy in those days, and of course they couldn't afford a car. Their dad was at sea most of the time, and the boys' mother Emelia took care of them. They attended the Lutheran church when my father was in town, and he carried the faith with him his entire life. The farmhouse was so far from the center of town that the community could not afford to extend a power line or "lightbulb pole" as they called them. So the Hansens lived without electricity.

The winters in Karmoy were terrible. They burned kerosene for light and for cooking, and heated the place with a woodstove. If one of the brothers wanted to leave the room, he took the lantern to find his way. The rest of the family would sit in the darkness until he returned. There was no running water, just a frozen outhouse to rush out to in bitter-cold snowstorms. They collected the rainwater off the roof in barrels, and stored it in a little tank for the rest of the year. During the war it was difficult to get groceries and commodities, so they ate mostly local foods, lamb and pork sausage, salted cod, and herring—lots and lots of herring.

Sverre was an intelligent boy with a photographic memory and an active imagination. As the war raged on, he heard the stories of brave Norwegian soldiers who, disguised as fishermen, were smuggling Home Front agents back and forth between Norway and the United Kingdom. The operation ran between Norway's western coast and the Shetland Islands, and was nicknamed the Shetland Bus. The North Sea was heavily patrolled by German U-boats and destroyers that would fire on any military ship. So the Shetland Bus depended on small wooden fishing boats; the sailors wore fishermen's clothes in case they were inspected. They carried light machine guns and smuggled weapons, radios, and other equipment to the underground resistance in Norway. After losing a few boats and a number of men to German patrollers, the Shetland Bus switched over to newer, faster "submarine chasers," 110-foot diesel-powered ships capable of running at 22 knots—about 25 miles per hour.

Of the 198 missions run by the Shetland Bus, 52 were captained by Leif Larsen, a daring sailor whom history would remember as

Shetlands Larsen. In the fall of 1942, after attacking a German battle-ship from the tiny wooden fishing vessel *Arthur,* Larsen became a na-tional folk hero, half patriotic freedom fighter and half gutsy rebel underdog—a cross between George Washington and Robin Hood. Larsen would go on to become the most decorated navy man of World War II—from any nation. Young Sverre drank up the stories of Larsen like they were a magic potion. Here was a man who was master of the sea, a brave war hero, and kind of a fisherman to boot. Listening to the stories of Shetlands Larsen, Sverre imagined for himself a life of sea-faring, adventure, and heroism. He would count Larsen as his hero for the rest of his life.

There was a lake in Karmoy that everyone skated on during win-ter. It seemed a good testing ground for young Sverre's bravery. One day he skated all the way across the lake with a friend who lived on the other side. It hadn't been too cold, and right in the middle the boys noticed a hole, fifty yards across, where the ice had thawed. It was easy enough to avoid. Later that afternoon Sverre had to get back across. Darkness was falling and he knew it would be harder to see the hole in the lake, but surely he wasn't going to walk around the thing. He was going to be brave, just like Shetlands Larsen. So he set off alone. Sure enough, out there by himself, no more than twelve years old, Sverre broke through the ice into the freezing water. With the last of his energy he threw a leg over the ice. Luckily his skate had a long blade, and he hooked his heel and pulled himself up.

A few hours later his mother wondered why he wasn't home and sent Karl out to find him. Someone told Karl that his brother was over at a friend's house, drying out. Apparently he was worried he'd get in trouble and had decided not to come home quite yet. Maybe that's a Hansen trait: getting yourself stupidly into trouble, just barely saving your ass, then shutting your mouth about it so you don't catch hell.

In 1948 the Hansens moved into a new house, about 150 yards from the beach. When Sigurd brought his sons to the cottage for the first time he led them directly to the light switch.

"Karl, Sverre, you ought to try this," he said proudly. "The light comes on by itself!"

Running water and on-demand electricity notwithstanding, Norway was ravaged by the war, and staggered on afterward like a Third World country. Even with the Marshall Plan pumping in money from America, it remained hard to get food and durable goods. Without the herring industry, many Norwegians in Karmoy would have starved.

The boys helped their father when they could—hanging up the cotton nets on Saturday night so they'd have a chance to air-dry before Monday morning. At age thirteen Sverre took a job at Vassvaag's sausage shop as an apprentice for the old butcher, who always wore wooden clogs that he dragged across the floor with a limp. Sverre worked six days a week for a salary of thirty-five crowns—about five dollars. In addition to skills like butchering carcasses and stuffing *fare polse,* a dried and smoked lamb sausage, Sverre became an expert at throwing knives. He liked to fling them across the room so they stuck in the wall. His favorite trick was to hurl a knife at someone's back as he left the room, so that it impaled itself on the doorframe, shuddering with momentum and scaring the hell out of the person. It gained him a nickname that followed him to Seattle and Alaska: *kniven stikker* or knife sticker. Maybe that's where I inherited my love of throwing lawn darts at my brothers.

Sverre also operated the meat grinder. In addition to producing sausage for human consumption, Mr. Vassvaag supplied junk meat to the local fox farms. One morning Sverre was shoving the guts and bones into the big grinder when the machine grabbed his finger. The thing could have easily gobbled up his whole arm. Luckily he pulled back, but not before the last knuckle of his ring finger had been chopped off.

Sverre's other duties included making deliveries and pickups around town on a bicycle. Later Karl would get the same job. Karl was supposed to make the deliveries on his bike, but he coveted Mr. Vassvaag's three-wheeled motorcycle. The butcher caught the boy eyeing the vehicle.

"Don't take that motorbike!" he called.

"No, no," said Karl. "Of course I won't."

As the boss watched, Karl balanced the wooden box of meats on the back rack of the bicycle and began fastening it with a rope. Instead of actually tying it, though, he just laid the rope there so it *looked* like he was tying it. Satisfied that Karl had done right, Vassvaag left the window and returned to work. Karl snatched the box, threw it on the three-wheeler, and *vroooomm!*

"Oh, you son of bitch!" the butcher yelled after him.

Same thing every time, until finally he just let the kid run the motorbike.

When the butcher purchased a live animal for slaughter, he sent one of the Hansen brothers to retrieve it. But it wasn't that simple—the boys were expected to kill and bleed the thing before packing it. Once when Karl was thirteen he was dispatched on such an errand to collect a veal calf. The animal had never seen daylight and when the owner brought it outside it went wild, running and kicking. Young Karl wielded his axe. With the farmer looking on, he swung for the head, but missed most of the time as it thrashed about. Finally he landed enough blows to knock the creature out. Then he unsheathed a brand-new knife to bleed it. If it wasn't bled immediately, the meat would be ruined. Karl stuck him. He plunged the knife into its throat, but when he stood up, only the handle was in his hand. The blade had snapped off and remained in the throat.

Jesus Christ, Karl thought. *Not a drop of blood.*

In a panic, he asked the seller for a knife. The man ran down to his basement and came back with a dirty old knife that hadn't seen a sharpening stone in years. It was dull. Karl started carving and eventually opened the throat. He got the job done, hoisted the carcass onto his motorbike, and headed back to the shop.

In another time and place, Sverre might have become anything—a doctor, an architect, or an engineer. But in Karmoy in the fifties, there weren't too many options. When a boy turned fourteen he was confirmed by the church and considered a man. Then his parents expected him to find work and make his own way. Some young men pursued school, but Sverre didn't have the money or connections to take that path.

My dad quit school after the seventh grade and began fishing. I'm still amazed that in America a lot of people think that if you don't go to college, you've got no future, and they look down on you. My dad was one of the smartest men I ever knew, and one of the most successful—and he did it on seven years of school. So I just never believed that a college education was all it's cracked up to be.

Herring was the mainstay in Norway, and in Karmoy. There was a breakwater at the harbor, and it was so thick with herring boats that you could literally walk across them from one side to the other. Sometimes the money was decent. Other times, like on a herring run to Iceland, they filled up the tanks and steamed into the processor only to find its owners flat broke. The processor wouldn't buy a single fish. It was almost the same thing that had happened to my grandfather a generation before. Sverre's skipper had no choice but to dump the fish and steam back to Karmoy, three weeks of hard work for nothing. And deckhands like Dad weren't paid a wage—they took a percentage of the sale. No sale meant no money.

Fishing was harder then than it is now. The raingear wasn't even waterproof. In those days, the women made giant wool clothing by hand. They soaked it in salt water, and when dried in the sun or by the stove, the garments shrank. The women repeated this until the jackets and gloves were so tightly knit that they repelled water—at least for a while. That's what Sverre wore. "Hand shoes" is what the old-timers still call those gloves.

Karl went to sea, too, with a job as a galley boy on the *Vigra* with his father—this was before it sank. He cleaned dishes, pots, and pans. He was paid one hundred crowns a week, about fifteen dollars. Nineteen fifty-six was the best herring year ever.

Then the herring fishery collapsed. Nineteen fifty-seven was terrible. They almost stopped fishing altogether in Karmoy. In those days, the best a fishermen could earn in a year was ten thousand crowns. Then they heard that in America the manshare was ten thousand *dollars* per year. The dollar was worth seven crowns. *Holy Moses,* they thought. *That's seven years' wages! We're in the wrong place.* That's what brought them over to America.

A lot of people were immigrating. Many went to the East Coast, fishing scallops out of New Bedford and Gloucester, Massachusetts. They would take passage to New York on a steamer called the *Stavanger Fjord,* which departed from the port city of Stavanger, a short ferry ride south of Karmoy. As it left Stavanger and chugged toward the Atlantic for an eight-day crossing, the big steamship ran close to the coast of Karmoy and it blew its horn, like a local bus hauling the island's young men to America. The boys came to the rail to wave farewell. Some never returned. In 1962, the scallop boat *Midnight Sun*—with ten Karmoy boys aboard—went down in a nor'easter off New Bedford. All hands were lost.

Dad decided to strike out for America. He had heard that some of the New Bedford fishermen were heading west to Seattle, where the fishing industry was booming. In those days you needed a sponsor—someone already in the United States who would deposit $500 in a bank account, sort of an insurance policy. That way, if the new immigrant royally screwed up, there would be some money on hand to ship him home. In 1958, Sverre's Uncle Jorgen—Sigurd's brother, who lived and fished in Seattle—agreed to front the money.

Sverre boarded a plane to Copenhagen—by then airplanes had made passenger ships like the *Stavanger Fjord* obsolete—then Reykjavik, then Los Angeles, and finally SeaTac International Airport. As the plane circled for landing he pressed his face to the window for his first look at Mount Rainier, the snow-covered volcano rising above the clouds. Out to the west, the white peaks of the Olympics blasted out of the calm green waters of Puget Sound, a sight not unlike the glacial-carved fjords back in Norway. Tugboats, freighters, ferries, and trawlers chugged across the harbor, crisscrossing the sound to the islands and the peninsula. Everywhere beneath him an ocean mist rose from the thick stands of evergreens. This was a not a bad place for a sailor. He had just turned twenty. The world was suddenly a large and exciting place. Who knew what he might find in America?

Sverre's aunt and uncle met him at the airport and delivered him to their home, a pleasant little house on 78th Street near Third Avenue. It was a quiet, working-class neighborhood of clapboard cottages, small lawns, and budding fruit trees. His uncle led him to the bedroom where he would stay and looked around proudly at the life he'd made in America.

"Welcome to Ballard, Sverre."

Dad shortly after his arrival in Ballard. *(Courtesy of the Hansen Family)*

My story is my brothers' story, and theirs is mine. So before I go too far telling about my life, and my dad's life, I need to catch you up on my brothers, Norman and Edgar.

Norman started fishing when he was fifteen, on a salmon boat in Bristol Bay, with one of the Old Man's Karmoy friends, Pete Haugen. He worked a few seasons on the *Northwestern*. His first was blue crab at St. Matthew. The weather was horrible, but he had faith in the boat. He loved it. At first he started to get queasy. He thought he was going to puke so he went down to the bathroom. He stood there. He told himself, *I ain't gonna do it*. Ever since then, it's never happened. Edgar throws up on his first day out, just to get it out of his system. A lot of people are like that. I got it out of my system as a teenager. But Norman has never been seasick. He's also the only one of the brothers who doesn't smoke. He chewed Copenhagen for years and recently gave it up.

With all the money he was making, Norman bought a 1970 RS Camaro and customized it himself. His real passion was for the hi-fi. He pulled the backseat to make room for a two-thousand-watt amp and thirty-six speakers: two 18-inch woofers, three sixteens, four twelves, four eights, a bunch of sixes, and a dozen tweeters. He poured fourteen grand into that stereo and went cruising though the neighborhood. You could hear him a mile away. This was the early eighties, when car stereo technology was cutting edge. Norman took the Camaro to the Northwest Auto Sound Contest—and got second place. The vehicle that beat him was a Blazer with six thousand watts. The owner was a drug dealer, but Norm actually worked for his money. For him it wasn't about having a cool car, or even a great stereo. He just got obsessed with a project, with learning how a system worked, and once he began, he couldn't stop until it was perfect.

By the time Norman was nineteen he had a serious girlfriend. She said, Make a choice: the boat or me. So he decided to get a normal job to be with her. He'd seen a lot of marriages fail, and he knew how hard fishing was on a relationship. He didn't want to do that. He chose to have a family life.

Norman enrolled in Shoreline College and got a degree in automotive repair. Dad was incredibly proud—Norman was the first member of our family to get a college degree. Still is, come to think of it.

Norman was one of the top students and after graduation the teacher made a phone call to Bellevue Toyota and said he had a guy who'd be good for the oil rack. Norman started as a lube technician and worked his way up from there.

Meanwhile, the relationship wasn't working. He and his girlfriend split up. One of Norman's friends from work was always talking about getting out of the city and moving over to Eastern Washington, to be in the country, around horses, hunting, and fishing. He moved to Yakima and a year later he gave Norman a call and said you gotta move over here, it's great living over here. Norman drove over one day, checked it out, and basically fell in love with the place. He transferred to the Toyota dealership over there.

Eastern Washington suited him well. Norman never liked the big city. He can't stand traffic jams. He also hates shopping. "It gives me anxiety attacks," he says. "If I need to go to the mall for a pair of pants, I call one of my friends' wives to go along, otherwise I just turn around as soon as I step inside."

Norman lived in the forest thirty miles out of town. He got into fly-fishing for a while; mastered all the techniques of tying his own flies. Then he moved on to pheasant hunting, but it scared the hell out of him—all the dogs stirring shit up and the birds flushing up, and having to figure if they're male or female while your buddies are blasting away. Next he tried goose hunting, but all he did was sit in a field and freeze himself to death. Then came duck hunting, but nothing ever came of that. He would fire away—*boom, boom, boom*—and end up on the ground trying not to get shot by the others. Norman hunted deer, elk, cougar, and bear. He got into hard-core jeeping and amateur bull-riding at the local rodeos. He also took up photography, but for more than a decade he was *not* a fisherman.

Then there's the baby of the family, the one who never misses a chance to make some smart-ass comment behind my back—usually with the camera running. Edgar. Like Uncle Karl, Edgar was never going to be satisfied to sit in the shadow of an older brother. He has always done things his own way.

One summer Edgar had just got back from a long opilio (snow crab) season and was driving around Seattle in his truck when he saw a black Saleen Mustang convertible with the roll bar and everything. He just fell in love with it. He was about twenty years old. He drove straight to the Ford dealer and there was a white convertible sitting on the turntable. Turns out it had belonged to Ken Griffey Jr. The baseball star had had it for about three thousand miles.

Edgar had just flown in from Alaska. He had a full beard and long hair. He smelled of diesel fuel, sweat, and rotten fish. He hadn't taken a shower. Nobody wanted anything to do with him. He looked like a homeless person.

He looked at the Mustang and said, "I want to buy that car."

The salesman just chuckled. "Yeah. Don't we all?"

It wasn't the first time Edgar had been treated like that. He had started fishing when he was seventeen. Sophomore year the Old Man asked Edgar if he wanted to go up for the summer. They needed a hand up there so he said alright, I'll give it a shot. He left before school ended, and the teachers gave him homework to do while he was gone. So he packed his schoolbooks and went up, not knowing what he was getting himself into.

This was the middle of an opilio season. Back then you fished opies for six or seven months straight, from January all the way to August. So he met up with us in June. Edgar had helped doing repairs on the boat in the summers, rigging pots and painting, but that gave no clue as to what actually went on.

When he got off the plane at Dutch, it was a whole different world. No one in the family had told him what to bring. He didn't have gloves, raingear, or sweatshirts.

"This was a tough-love family," Edgar says. "No hugging, kissing, all that crap. It was all about respect. You had to earn your respect, and I had earned none from the way I was acting."

It was a culture shock. He was getting yelled at all day: *Get this! Get that! Get that bait onboard!* There were four guys onboard and they were all seasoned older men. Brad Parker, Pete Evanson, Nick Balahoski, and Mark Peterson. The best of the best.

"I was supposed to be the bait boy," he remembers, "but I was just lost. I was used to driving my car around and sitting home playing video games and drinking with the boys."

It was bad. We finally left port and he got sick. He was in his bunk for damn near three days. It's expected for greenhorns to get seasick, but usually the guy will at least get up and try to work. "I was just being lazy," Edgar says. "I didn't know any better. I was being paid a hundred dollars a day. So I just lay there."

After three days rocking in his bunk, he still couldn't snap out of it. The crew was out there sorting crab and finally I couldn't let it go on any longer. I went down to Edgar's bunk. "That's it," I said. "You're done." I grabbed him by his hair and yanked him out of bed. "Get outside. You'll feel better. I promise."

Edgar finally got dressed. All he had was jeans and a T-shirt. He had to borrow raingear. He got out on deck and was just miserable. He was chopping bait. He had one bait jug and one fish to cut, per pot. "I was worthless the whole time," he says. "I never snapped out of it. It took over a week just to stop puking. Just went on and on. The guys were looking at me like, *The owner's kid? This is what we get?*"

Even after, for two months, he still didn't get it.

"Something new was coming at you everyday," he says. "Bad weather, or bad crab, or dead loss. And then we had some bad fishing, and there was a point where I actually made more money than the crew. Guys were pissed, because I wasn't holding up my end of the bargain. I was behind on every pot. Lazy. Slow. I had no work ethic. I didn't know what work was. I'd had jobs down in Seattle before, running meat grinders in grocery stores, washing dishes in restaurants, but that was all."

He finished the summer. When he came home the Old Man said, "What did you think?"

"I hated it."

Dad just laughed. "It'll get better next time," he said. Sverre wouldn't have minded if Edgar didn't want to fish, but he did want to see what his kids were made of.

One thing Edgar had managed to do that summer was all the homework the continuation high school gave him. He turned it all in.

"So when can I start?" he asked the principal.

The guy took one look at Edgar and said, "Sorry, we don't have room for you."

"What do you mean?"

He said they were overenrolled. Edgar had received his check from fishing, which wasn't all that big, but he'd heard how much the full-share guys were making. "So I said, fine. Fuck you then. Who needs you?"

A couple months went by, and he went back to fishing. He had no choice. He probably could have gone to a community college, and gotten a GED, but he thought, *You go to high school to get a diploma, then you go to college to get a degree, so then you can get a good-paying job. Shit. Why not eliminate the middle man? Go straight to where the money's at. No brain, a strong back: that's all you need.*

So Edgar went back to the *Northwestern*. "I was living in misery. I hated being there. Sometimes I would literally cry myself to sleep. I was in so much pain and agony. My hands were just locked up and bleeding from split fingers. I didn't want to be there. I didn't know why I was there. But at the end of each season, after being home for just a few days, I thought, *You know what? That wasn't so bad.*"

"It's an insane mentality that runs in the blood," Edgar says. "It was in the blood from the beginning, but it took me a little longer to get it."

After about a year, even though the Old Man owned the boat, Edgar was still standing at the sorting table making only a half share. He was leaning his elbow on the table, pitching crab over his shoulder, one at a time. "I was only willing to work with one hand," he says. "And suddenly something whacks me on the side of the head, really hard. It's a flying crab. I look up and I got four guys staring me down with the look of the devil. Pete Evanson had just brained me with a king crab."

"You got two hands," Pete said. "They're not broken. So use them."

Edgar put his head down and tears started falling. He started sorting. It didn't take much longer after that for his attitude to change.

"Something just snapped inside of me. I don't remember exactly when or where. But after being home, I realized that I missed it. And the next time I went up I was full-on, gung ho. I worked my ass off. I learned whatever I could from the older deckhands, because obviously this was what I was going to do for the rest of my life."

He worked hard and played hard. Edgar says he could probably own five houses by now if he had lived like a normal person. "But I was into booze and drugs and the rest."

They would be doing gear work, rigging pots on a hot sunny day, then look around, and someone would ask, "You want to go to Mexico?"

"Sure, what the hell?"

They'd make a phone call and book a ticket. Bam! They were gone, down to Cancún where they'd blow five grand.

We were a hidden little community. Nobody knew we were making the huge money we were making. Nobody knew how hard we were working for it. The bars knew it, though. They knew it because when the fishermen were home the tabs were unbelievable, and so were the tips. The bars made their living off of the fishermen.

One time in Dutch Harbor, Edgar got so desperate to play guitar that he went from boat to boat trying to buy one. He wanted to take it on the next trip. Yet nobody who had one wanted to sell. Finally he found some Russian on a big freighter with a pawnshop guitar worth about fifty bucks. Edgar paid him seven hundred cash.

Edgar couldn't figure out how to pay his taxes. We were paid once a year, and it was a 30 percent chunk. So he was paying between thirty and forty-five thousand dollars in taxes. Edgar had no write-offs; he didn't know how that worked. When tax day came around, all his money was spent. "So I had to borrow against the next season from the Old Man," he says.

One year we'd done really well, made a lot of money on king crab. Christmastime rolled around, with January coming up, and we were getting ready for opies. We were supposed to leave in a couple days. Edgar had fifty thousand dollars in the bank. He was sitting around a buddy's house thinking it would be kind of nice to get his high school diploma. First he had to call up my dad and tell him he didn't want to go. "I dreaded that call," he says. "That was the second hardest phone call I ever had to make."

When Dad finally got on the line, he said it wasn't a big deal. "You gotta do what you gotta do," Dad said.

So Edgar stayed in Seattle. By the end of the month, he hadn't gone to school, and all that money was gone. "Women and wine and song were a big part of it," he says. "I was flat broke. I had car payments, rent, and then taxes coming due. The next phone call was the hardest: I had to tell my dad I didn't go back to school, I was broke, and I needed to either borrow money or have my job back. I should have just gone to

Edgar in the engine room of the *Northwestern*. *(Courtesy of EVOL)*

the bank for a loan. It was worse going through him. He gave me a look that made me feel puny. He was a powerful, powerful man."

Even then the Old Man wasn't too mad. He gave Edgar his job back and Edgar went up and finished opie season. And he loaned Edgar the money for taxes—with interest.

That winter when Edgar saw the Mustang for sale, he had already cashed his paycheck. So he went directly from the bank to the Ford dealer, stinking like fish and fuel. He was wearing his *Northwestern* jacket, but that didn't mean anything to the guy.

"No, really," Edgar told him. "I want that car. I want it now."

"Yeah, yeah. We've got some used cars in the back."

"No, seriously."

So he pulled out the wad. Twenty grand. Edgar laid it on the table.

"You don't understand, I wanna drive that car home today. Excuse the looks, but I am a human being. And I have money."

"Let me help you out, my friend," said the salesman, warming up quickly. "Let me go talk to the manager, and see what I can do."

"Whatever," Edgar said. "Sign on the dotted line and give me the car. I want to go."

So they worked out a deal. It took about twenty minutes.

"I didn't barter," Edgar says. "I didn't even know what bartering was. I'd never owned a new car before. I drove it off the lot. That was that."

3

BALLARD, AMERICA

Captain Sverre hauled himself up the ladder and yelled to wake the crew. The other two deckhands came stumbling from their bunks in slippers, dungarees, and T-shirts. Now the four men raced onto the deck and began unrolling the hose. The bitter arctic wind howled, but with the surge of adrenaline Sverre was not cold. He knew he only had a few minutes to contain the blaze. The old diesel-soaked boat was a tinderbox.

The men looked slightly comical wielding the hose like firemen yet dressed in slippers, but they had no choice. The boat heaved as waves crashed over the rail and soaked the men to their knees in frigid seawater.

How were they going to fight this fire? They couldn't get belowdeck to the engine—it was already engulfed in flame. Sverre had an idea. Along the base of the cabin were a series of vents leading up from the engine room.

"Are the pumps working?" he yelled.

Krist hollered that they were. As long as the engine was running, they would have water pressure as seawater was pumped up from an intake valve in the hull. If the engine died—well, Sverre wasn't going to worry about that now. The hose was industrial black rubber, a bit thicker than a garden hose. It was intended for irrigating a fish-cleaning table and for scrubbing guts off the deck—not for firefighting.

"Open a vent!" he ordered. The men did as they were told and Sverre shoved the hose down it. This was going to be a mess—if they did put out this fire, they'd swamp the engine room and maybe sink the boat. Suddenly there wasn't a lot for the four men to do but stare at the hose stuffed down the vent. Was it working? They couldn't see the fire, so they couldn't tell if they were dousing it. With the steering dead, the wheelhouse empty, and all four men on deck, the little wooden boat was rudderless, tossed by the seas, riding sideways up the crest then sloughing into the trough. The wind howled.

Finally they pulled the hose and Sverre peered into the vent. No flames. No smoke. Could it really be that easy? *It was too good to be true*, Sverre thought. Or, at least, the alternative was too bad to consider.

"Open the next vent!" he called.

Krist pulled back the vent cover—and was greeted by a hot, licking flame.

"Give me the hose!" he yelled. The men brought the hose and stuffed it down the hot vent. Once again they waited, their sweat freezing solid on their shirts, but as soon as they doused one flame, another leapt from a different hatch. The wind howled and whipped up the blaze.

They didn't have time to play this game. Sverre had to consider what he'd been trying not to think about. Belowdeck was a bank of marine batteries, each like a small grenade ready to explode. There were oxygen tanks used for welding, which would ignite like artillery shells. There were also thousands of gallons of diesel fuel sloshing in the tanks. Captain Sverre didn't have the luxury of not snuffing this fire.

The *Foremost* was a floating bomb.

In 1984, I finally finished high school. All I wanted to do how was fish. Just before the king crab season, we were loading crab pots on the boat in Ballard. We brought them in on semi-trucks, stacked three or four high, backed them up to the boat, and the cranes hoisted them aboard. I was standing on the stack, trying to hook the crane to the top pot, but the stacked pots were jutting halfway over the end of the bed. The truck driver let go of the strap and the pot I was standing on started teetering. They were giving way beneath me. I dove forward to the next row. As I dove, the falling pots hit my ankle and flipped me over. I landed on the bed of the truck and crashed to the ground. The crab pots were bouncing all over the street—700 pounds each—and one landed on me. I busted open my head, cracked some teeth, and broke my ankle. I was out for the season.

When I recovered I went straight back to Alaska. I had had a high school girlfriend, but as soon as I got that ticket to Dutch Harbor, I broke it off. I just wanted to fish, and finally I got what I had wanted for years: I became a full-share deckhand on the *Northwestern*. We were gone nine to eleven months out of the year. We'd go up to Nome for red crab, then jump over to St. Matthew for blue crab, which after my first trip was still my favorite. Then we'd go for Bering Sea red crab, which is what you see on TV. Then we might go way out west and take a chance on Adak red crab. Starting in January we were back in Dutch Harbor for opilio crab. In the early eighties, if they hit it right, crewmembers were making up to $150,000 per year.

That's when I made some of my best friends. In addition to the

Norwegian kids who'd grown up in fishing families, we found some others in the neighborhood. Mark Peterson had gone to high school with Norm and me. "I moved from a working-class neighborhood in Massachusetts," he remembers, "where there were three families living in one house. My parents divorced, so we moved to this area. And I see all these kids driving around in these fancy cars, and I think, *What a bunch of spoiled little brats.* Then I found out they fished for a living. *What do you mean you fish for a living? Who fishes?* So I found out they went up in the summers and did some salmon fishing and then they also went on the crab boats and made enough money to buy this stuff. Well, I had to get in on that."

Peterson asked Norm, "Is your dad hiring?"

"I don't know," said Norm. "Why don't you ask him?"

So every time Peterson came to our house, he'd go find Sverre. Every time.

"You guys hiring?"

"No."

End of conversation.

Peterson kept asking. Finally we hired him to go down to the boat in Ballard and help paint it and chip the rust. He installed a washer and dryer with us, oddball stuff like that. "They paid me ten dollars per hour under the table," says Mark, "and that was fantastic, because I was working for $3.35 per hour at the gas station." He was a strong, compact guy, built like a wrestler. He kept asking for a crew job, but of course to get a job back then, somebody pretty much had to die.

So Peterson graduated high school, went off to the army, basic training and all that. He went into the reserves, was gone for a good six or eight months, then came home for Christmas and had community college lined up for the winter semester. In the meantime he was back at the gas station. Then one day I happened to pull in to get gas.

"Hey, Peterson, you still want to go fishing?"

This was a Friday afternoon.

"When?" he said.

"Monday."

"Yeah."

Peterson quit his job, called the school and canceled, and put his scholarship on hold. He and Norm went down to Seattle Ship Supply and bought a bunch of sweatshirts, oil skins, and this and that. Then we took off and went fishing.

By the time I was working full-time on my father's boat, he was seldom working as skipper, choosing instead to manage the business from home in Seattle. Tormod Kristensen was usually running the *Northwestern*. My brothers and I learned the ropes from him. Tormod was a great skipper and a mentor to us all. He became captain of the *Sea Star* about the same year my dad became captain of the *Foremost*, and did very well on many boats over the years. "Of all the guys I worked with in Alaska, Tormod was hands down the best," says Mark Peterson.

When Tormod was skipper, we always slept from midnight to six. Banking hours. "You can't work safe if you don't get sleep," Tormod said. During sleeping hours, everyone took an hour-and-fifteen-minute watch, including the captain. "It was like clockwork," says Peterson, "and we outfished almost everybody." In the morning we'd hear him climb down the stairs and flick the light switch. "Ja, okay. A couple miles," he'd say, then sneak back up to the wheelhouse, almost like he was afraid to interrupt our sleep. What he meant was that we had a half hour to get dressed, eat, and get on deck. Otherwise he'd be pissed. Not once, though, did he ever get mad enough to yell.

He was soft-spoken, partly because his English was never great, but mostly because of his shy Norwegian temperament. "It's not like my wife's cooking," he liked to say whenever one of us cooked up a meal.

Tormod understood weather better than the weather service. He'd look at the barometer. "Ja, it's gonna blow." It always did, if he predicted it. Whenever Tormod climbed down to the galley he poured himself a cup of coffee and took two cubes of sugar from the dish. He dipped one in the coffee, took a nibble, then threw the other half in the trash can. He took the other sugar cube upstairs to the wheelhouse. By the end of the day, a mound of melted sugar had hardened like glass in the ashtray.

I can only remember one time when Tormod really lost his temper. He loved birds. One time he caught us capturing seagulls to use as bait. He lost it. "Goddamn it, man! If I see that again, you're all fired."

A lot of fishermen didn't want to work on Norwegian boats because the skippers were so crazy about keeping them clean. One of them wouldn't bring his boat in to port until we had scrubbed every square inch. He circled the harbor while we got on our hands and knees and scrubbed and hosed until our fingers were raw.

"It's clean now," we said.

"Not clean," he replied. The skipper dropped the anchor and we scrubbed everything down again.

Tormod never bragged, but he was always one of the best fishermen, a highliner. Once my Old Man was in the bar talking about how he'd extended the *Northwestern* to hold more pots. "It's not how many pots you pull, Sverre," Tormod said quietly, with a little smile. "It's how many crabs you have in the pot."

You worked so hard for Tormod that your fingers would crack and lock into a claw. You couldn't straighten them. You couldn't shake down the crabs because your hands hurt so bad. At night we smeared our hands with bag balm, stuffed them in cotton gloves, and put them under the pillow while we slept to flatten them out. Then we woke up and guzzled ibuprofen. We worked a six-month opie season without going home. We had a *Sports Illustrated* calendar on the wall and we didn't follow it day by day, or week by week. One morning Norman stumbled out of bed and asked, "What month is it?"

We would get dead-dog tired. During wheel watch, men would sleep with their eyes open. We called it sleep-watching. Your eyes felt like sandpaper, like they were on fire. I remember dreaming with my eyes open. I thought I was on a boat called the *North Command*. I looked at the radar screen and saw two dots, but in the glare of the window I saw hundreds of lights. I jolted awake, thinking there were fifty boats ahead, when it was really just the hallucination caused by exhaustion.

We did whatever we could to pass the time and break the monotony. We invented the Button Game. Everyone would gang up on one person and "push his buttons"—needle and tease him to see if he'd break. We'd haze each other all day just to end the monotony of the work. If Norm was the target, we'd blow on his food—he was so disgusted that he'd push away the plate without taking a bite. When Peterson was a greenhorn, just after he'd completed basic training, he said, "There's nothing you could do that my drill sergeants haven't done." So Johan and I decided to freeze him off the boat—lock the cabin door and see how long he could stand the subzero cold. To cope, Peterson just started doing pushups and chin-ups to keep warm. That blew us away.

Sometimes the guys would haze me. One time, I was throwing the hook, and just as I let go, Johan stepped on the line to irritate me, and the hook fell short and plunked into the water. I hauled it in and threw it again, and he stepped on it again. Finally, Tormod had to turn around the boat because we missed the buoy. I had a short fuse. The fuse lit, and I chased Johan and threw him into the bait freezer door. We were laughing even as I flattened him like a pancake. At the end of the day we still worked hard and were still good friends and shipmates. That's how things were.

We had other ways of pushing each other's buttons. One of the favorite meals was Norwegian meatballs that I'd make for the crew. They were pretty damn good so when I was on deck the crew would sneak into the galley one at a time and plunder them. They were eighteen-year-old kids, hungry as hyenas, but when the captain wanted his meal and there were no meatballs left, he'd blame me. It was embarrassing because it made me look bad. Dinner was my responsibility, and I took pride in my cooking—a skill I'd learned from the older men. So I decided the best solution was to hide the meatballs. This only made the others more determined to find them. There's not too many places to hide a platter of meatballs on a ship. I laid them on a paper towel and tucked them inside the machinery of the stove, under the burner, but the bastards found them, too.

It's really hard to wake me up, especially after a few days without sleep. When it was my turn for a wheel watch, the crew would pound on my chest, yell in my ear, and still I wouldn't wake up. Peterson once solved this by whispering in my ear, "Sig, your roast is burning." Still asleep, I jumped out of my bunk, ran to the galley, and pulled open the oven door. Cold and empty.

"I'm not cooking a fucking roast!" I shouted, half-asleep.

"But it got you out of bed," he replied. He got a kick out of that.

Sometimes the hazing could get a bit too personal. One morning, the crew simply could not wake me. They decided to drop a string of pots without me. Tormod asked where I was and they said they didn't need me. I think I slept for six hours. I woke up and I knew something was wrong. The worst thing you can do on a crab boat is not pull your weight. I dressed in my raingear and ran out on deck. All the pots were unloaded. I couldn't believe it. The other three just sort of glared at me. Peterson stepped up. "That's all right, Sig," he said. "You don't have to work. It's your daddy's boat."

I punched him square in the face. He came back at me and the others broke it up. All the fighting made us close, like brothers. We're still friends, twenty years later.

"One of the proudest moments of my life," says Mark Peterson, "was steaming into the Ballard Locks after my first season on the *Northwestern*, knowing I just did seven months on a crab boat in Alaska. It was sunny out and we cleaned up the boat because we wanted it to look so good when we got home. I'll never forget that."

We were growing up fast, in a world with lots of money and few rules. There could be no more colorful a backdrop for this coming-of-age than the Aleutian Islands. Back then, Dutch Harbor had a Wild West flavor. The roads were unpaved and things could sometimes get unruly. During crab season—also known as the Derby—all the fishermen would race one another for their share of the catch. Dutch Harbor was crazy the week leading up to opening day. Almost three hundred boats would arrive to try and get a piece of the big money. The Elbow Room would be packed. They had live music. Inside, the ratio of men to women was about 10 to 1, so you were lucky to find a barmaid or a cabdriver who would take care of you when you were in port.

Fishermen in their twenties were making a lot of money. Some would get paid a big settlement in cash and forget about Uncle Sam. If they got a letter from the IRS, they would just jump from boat to boat and hope they stayed ahead of the tax man. Most times the government tracked them down.

There was a lot booze and drugs around. It was the mid-eighties, the height of the cocaine craze across the nation, and Dutch Harbor was not immune. The stuff was everywhere—you didn't even have to look for it. The men of my father's generation disapproved, but the younger guys were spending a lot of money on it. I was never much into drugs. I did dabble, but one morning when I was twenty-two I told myself this is not going to work. I flushed what I had down the toilet. That was the end of it.

Fueled by the drugs and booze, arguments broke out all the time. If you had a good trip, you could come into town with bragging rights. People always heard what other boats were doing. Some guys would brag too much, others might get jealous, and a brawl might break out. It happened all the time. Around 1980, after they built the bridge from Unalaska to Dutch Harbor, water taxis were replaced by car taxis. It was three dollars from point A to B. That was the most dangerous part of the whole trip: when you got guys from different boats in the same cab. Soon they were bragging and throwing insults. The next thing you knew fists were flying. The cabdriver was usually some tiny Filipino lady, and she didn't know what to do but pull over and wait for everything to sort itself out.

There was this big old monster of a guy—I'll call him Bad Bart, although that's not his real name. He had bushy hair and biceps like kegs. One night he was in the Elbow Room when my friend and I went in. We'd had a few, and I bumped into Bad Bart without even knowing it. I didn't care.

"He's going to kill you," said my friend, who began to panic. "You didn't say you were sorry." He grabbed me and pulled me outside. Taxis—if you could call them that—were lined up out front. They were actually mud-splattered, smoke-riddled rusted old vans. My friend pushed me into a stinking vehicle. "We better go," he said.

Suddenly, Bad Bart barreled into the cab. I'm a little guy—all I could do was grab him by the hair and throw his head down on the taxi floor. I had wooden clogs on—they used to call them skipper slippers—and I just started kicking his face in with all I had. Then another guy grabbed him and threw him out, and jumped in the cab. We took off. The other guy was here to help me. He put me up at a room in the UniSea. "Just stay here," he said. Later we went back to the boat and Bad Bart was hiding there, with a wrench in his hand, just waiting. So we went back to the hotel until he left. Let's put it this way: It was a long time before I showed my face in the Elbow Room.

Another night some guy came back from the bar with us. We didn't really know him. He brought some coke, which pissed us off. Our crew could do whatever they wanted off the boat, but onboard was like a sanctuary—Norwegian rules. The guy proceeded to get more and more wasted. At one point he staggered into the galley and started pissing in the stove. Edgar followed him over and kicked him in the ass. All hell broke loose. Edgar hit him over the head with a bottle and slit him from eyeball to chin. There was total mayhem with blood all over the walls. The next morning Mark Peterson found the guy's wallet and coat in the boat, so he went down the docks looking to return it. Finally Peterson found him, with a row of stitches down his face. He didn't even remember what happened. "Was I still fighting when I went down?" he asked. Peterson told him he was. That's all the guy wanted to know.

Just as often, however, I'd get a good beating. I'd have some drinks and shoot off my big mouth, then get punched in the face. One time I went on board the ship docked next to ours and started giving them shit. I told them their boat was a piece of crap, a floating coffin. My shipmates could see where this was heading, and they tried to pull me out of there, but I wouldn't budge. So they left without me. Me and this other guy kept arguing. Finally we decided to go settle it on the dock. He was much bigger than me. I got a couple of licks in, but then he let loose with a big haymaker, spun me around and laid me out. That's all I remember. In the morning they found me bloodied and black-eyed, passed out on a stack of pallets on the dock. I had a concussion. It's just luck I didn't roll off into the water.

It really was just like the Wild West. Even law enforcement was sometimes over the top. As the people became more and more unruly, the cops became more vicious. One time a friend was in the bar and trouble started. The cops just took him up to the top of the hill and beat the crap out of him—better than having to fill out any paperwork.

Though not as wild as Dutch, Akutan was even more of a frontier— the whole island had only one cop. One night Mark Peterson was playing pool at the Roadhouse. The constable was just sitting at the bar drinking a can of beer—which was all they served. Mark won the game and the guy he beat hurled a pool ball across the table, point-blank. It whizzed right past Mark's head. Mark ran after the guy, caught him outside in the snow, and pummeled him. When he came back inside, the cop hadn't budged from his barstool. He looked up from his beer.

"Did you get him?"

"Yeah, I got him."

"Good," said the cop. He bought the boys from the *Northwestern* a round. Akutan justice. Later Peterson heard that the guy who got pummeled went back to his ship for his gun. Things could have gotten real ugly real quick.

There's another story about a crab crew that got busted for being rowdy, and the cop threw them in what was supposed to be a jail, but was actually more like a plywood box. By morning they were like, "We gotta get out of here. We're supposed to go fishing." They huddled together like a football team and knocked the wall down. They ran to

Matt Bradley in Akutan. *(Courtesy of EVOL)*

the dock and jumped on the boat just as it was leaving. The cop went screaming after them. The captain called in on the radio and said, "We'll get them back when the trip is over." They went out and finished their trip, then returned and did their time.

Amidst the lawlessness, we started to see evidence of the government regulation that lay in our future. Just about every fisherman has dropped a pot before Opening Day to sneak a peak at conditions. We call it "prospecting," and I've done it three times myself, although it is against the law. It's a way to be sure that as soon as the whistle blows, you're already on the crab. To combat this practice, Alaska Department of Fish and Game began inspecting tanks before the season to make sure they were empty and that we hadn't been fishing early. They conducted these searches not only in Dutch, but all across the fishery, in King Cove, Sandpoint, and St Paul. Fish and Game had planes in the air looking for illegal activity. They even used high-powered cameras to spy from a distance. The cameras were so powerful they could snap a picture from thousands of feet above and read what brand of cigarette you were smoking on deck. If they discovered a vessel breaking the law, they wouldn't stop it, they'd just let the crew fish. If the vessel caught a hundred grand worth of crab, the boat was fined a hundred grand. Another way they enforced their regulations was to count the crab a vessel brought to port. They did a random sample of one hundred crabs. If more than two or three were smaller than the legal limit, the crew was busted and heavily fined. Yet even the authorities weren't immune to the riches that we were earning. One time I urgently had to take a test with the Coast Guard, but the guy at the office told me there were no openings that week. I couldn't wait a week—I had to be back in Alaska in three days. The next morning I returned to his office and plopped a case of fresh king crab on his desk. He made an appointment for me right away.

Back then, we used to fly on Reeve Aleutian Airways. At the end of the season there was a mad rush to get the few available seats. Alcohol was forbidden onboard, but everyone carried a little bottle in their coat pocket. If they caught you, the bottle was simply taken away. What else could they do when you were in the air? Some fishermen figured they were going to get caught once, so they carried a second bottle. Back then we carried boxes of live crab home with us on the plane, and sometimes we'd let them out to crawl around the cabin, just to annoy the flight attendants.

Around this time I started to notice that my friends from high school were going to college. I wondered if I should do the same, so I asked my dad what he thought.

"Are you crazy?" he said. "You're making good money up there."

I was eighteen at the time and living at home. I visited a friend at the community college. I walked around and sat through a class.

"We should have a party," he said.

I brought him and a bunch of friends back to my parents' house. Mom and Dad were out of town and we trashed the place. It didn't take long for that news to get back to my folks. That one day in college would be my last.

So I stuck with fishing and became a lifer. My problem was that I never really had faith that there would be another season. I never took anything for granted. Maybe I was just paranoid. At any rate, the result of this attitude was that we put in more time than the average crew. For example, it used to be that only the ones who didn't make good money during the autumn king crab season would hang around for Adak red king crab season in the winter. Sometimes there would be almost 250 boats in the fall, but only 70 in the winter. Some crews wouldn't bother going out west to Adak because you couldn't make as much money there, but we went. Similarly, after opie season in 1986, everyone was fried and just wanted to go home. Instead, we went north to St. Matthew and fished blue crab and made a killing. We were one of the few boats there. Another year 170 boats went to St. Matthew. That was a huge jump from the year before, so the average catch declined. The following year, half the boats stayed home. We were there every year. Looking back now, I'm proud of it.

What a lot of people don't understand is that it's not always great money. We never went up and got rich by just fishing a couple weeks per year. We went every year and fished nine or ten months, damn near every crab season. We stuck with it through thick and thin, fished every season like it might be the last, and that's the key. The people who are there for glory will never last. That's not what it's about. Those people don't prove anything. Only persistence pays off.

With all the money flying around, it's understandable why someone would show up just to make a quick buck. For me, it's different. It's something that's been passed down from my father, and his father. The pride I have in my family's stories and traditions is much more important than the money.

It might make a good story to say that when Dad got to America he struck out on his own, pulled himself up by his bootstraps, and rose to the top all by himself. It's not quite true, because when Dad got to Seattle, other Norwegians had been building a community there for almost a century. In Ballard they had a town almost all to themselves, and in 1958, it may have been the most Norwegian place outside of Norway. To this day my wife and daughters and I dress in traditional Norwegian garb and march through Ballard for the Norwegian parade on May 17—Norway's Constitution Day, which is similar to America's Fourth of July. So when I tell Dad's story, and the Hansen saga, I have to back up and explain how all these Scandinavians got to Seattle in the first place and laid the groundwork for the likes of us.

Norsemen didn't stop crossing the Atlantic after Leif Erikson. The first group of Norwegians to officially immigrate to the United States of America left Stavanger—just across the water from Karmoy—in 1825. They are known today as the Sloopers, because they crossed the Atlantic on a wooden sloop. Descendents of the Sloopers think of themselves in the same way that the descendents of the *Mayflower* think of their past. "Sloopers," though, just doesn't have the same ring to it as "Pilgrims."

Forty-five passengers and seven crew boarded the *Restauration*, a wooden sailing ship thought to be about one hundred feet long and thirty-six feet across. About half of the seafarers were seeking religious freedom and the other half wanted land. Only about 4 percent of Norwegian land could be cultivated—the rest was ice, mountains, forests, and bogs—so by the early 1800s, as the population boomed, most of the land was taken. The 52 immigrants crammed themselves into 250 square feet of bunk space, and set sail from Stavanger in early July, with flags flying and guns saluting. To catch the trade winds, the *Restauration* sailed two thousand miles south, through the English Channel, along the coast of Portugal, and arrived at the Madeira Islands off the coast of Morocco after twenty-seven days. After a brief resupply they set a bearing for the West Indies. It was a rough, cramped, and stormy passage. On September 2, a baby girl was born onboard—perhaps the first true *norskamerikanere*, that is, Norwegian-American. Finally on October 9, ninety-eight days out of Stavanger, they reached New York. Those were some damn tough sailors. Most of the party eventually settled in Illinois.

The Sloopers' voyage began a century of Norwegian immigration, largely to the upper Midwest: Illinois, Wisconsin, Minnesota, and the Dakotas. Over the next hundred years, a full third of the population of Norway—about 800,000 people—came to North America. Due to its less stringent immigration policies, many entered through Canada, but most would end up in the States. The only country to send a larger portion of its countrymen to the United States was Ireland. Today, 2 percent of white Americans claim Norwegian ancestry. In the upper Midwest that percentage is about 15.

With the advances in naval technology, the sloops were replaced by steamships, making the crossing shorter and safer. Yet the passage was never without risk. In 1904, the SS *Norge* left Copenhagen en route to New York, its berths filled with immigrants. The ship sank at Rockall, off the coast of Scotland. One hundred sixty passengers were rescued, but 635 died, 225 of them Norwegian. It was the worst civilian maritime accident in history up until that time.

In those years, enough Scandinavians were living in the United States that the shipping lines could fill some return seats with passengers visiting home. In 1888, the SS *Geiser,* a 313-foot steel steamship set sail from New York to Kristiania (now Oslo). Though capable of carrying 700 passengers, on this return voyage it had just 110 plus crew. Among them was A. B. Wilse, who would later become one of Norway's most famous photographers. He wrote:

In midsummer we had finished our [railroad] work and were signed off in Minneapolis. These were not prosperous times in my line of work, and I decided that I would go back to Norway to spend the winter there, and return in the springtime.

I bought a ticket for Kristiania, and was scheduled to travel on the *Geiser,* one of the Thingvalia Line ships. With my trunk packed with my collection of souvenirs—Indian stuff among other things, actually all of my gear except my camera—I settled myself aboard the *Geiser,* which was lying in Hoboken. On a peaceful, quiet day in August [the] *Geiser* sailed off, with 110 passengers on board. Most of the passengers were people who were used to a roaming life, and we soon became like a huge family.

The first day passed, and we had a wonderful evening with calm seas, and a beautiful sundown. A lot of whales where tumbling about near the ship. On deck, there was dancing to the tunes of an accordion. Everything was happiness and joy—because of the moment, and the thought of where we were heading. On board there were not less than four captains going home on leave, there were also some teachers, businessmen, and farmers. As the evening got later, one by one they went to bed, full of joy and happiness over having had such a nice evening.

Hours later, near Sable Island off the Newfoundland banks, the pleasant voyage was violently interrupted. Here's how the ship's captain, Carl Moller, described it to a newspaper, "It was about three thirty o'clock when the first officer called me loudly. He was excited, and shouted, 'We are going to be run down!' I jumped from the sofa and ran out on the bridge in my nightshirt. I saw immediately lights of a big steamer on the starboard side."

The steamship bearing down on them was the SS *Thingvalla*, owned by the same line, running the opposite route. Of all the vast thousands of miles in the North Atlantic, these two ships somehow found one another. Captain Moller remembers,

With a tremendous crash, the bow of the approaching steamer struck us hard amidship, nearly at right angles to our keel. The blow took us just aft the main rigging, cut a quarter way through us, and made such a tremendous hole that I saw at once that we could not stay afloat.

I gave the orders at once to have the boats launched, to send up signal rockets and fire a gun. The confusion which followed, however, is beyond me. I cannot describe it. The boats on the bridge were launched, the starboard side one first. The man at the stern dropped the line, however, and the boat filled with water, and was swamped. Boat 2 on the port side was also launched, and she drifted away too far from the ship to be of any assistance. The only other boat launched was No. 8. The powder-room was flooded, so that no signals could be used.

The passengers now began to swarm up from below, and were completely panic-stricken. The confusion was awful. Men were struggling to get into the boats, and women and children were shrieking and screaming. I sang out to the lifeboatmen, "Look out for women and children first"; then I sang out below for every one to bring up life-preservers. There were between 600 and 700 of these on board. The panic was so great, however, that they did not pay much attention to them, but rushed on deck without them. The chief engineer, who was drowned, rushed down to the cabin for life-preservers for the passengers, and I never saw him again.

The panic was corroborated by *The New York Times* account from Mrs. Hilda Lind, "a comely Swede of 28" who lived in New York but was traveling with her three-year-old daughter and three-month-old son to visit her mother in Sweden. They were sleeping in their cabin when the ships collided. According to the *Times,*

One of the stewards told her [to] fly for the deck, that the ship was sinking. She asked one of the stewards to carry one of her children to the deck, but he made no reply and passed on. Several other men were in the passage, and to them she also appealed for help, but they paid no attention to her, all hurrying past to the companionway, up which they disappeared.

Mrs. Lind lifted the babe from the berth where it still lay fast asleep, and taking Ida's hand, started for the deck. A man brushed by them, knocking the little girl down, but did not stop to pick her up, although the mother begged him to do so. She knelt down, and, taking the child about the waist started to carry her to the deck. At the companionway a steward appeared, and the almost frantic woman asked him to carry her upstairs. He did not heed her, but scrambled upstairs. Mrs. Lind finally reached the deck, where she found an excited crowd, some climbing over the rail and others stamping the deck in frenzy. Women were screaming for help. She had been on deck but a few moments when the water swept over it, and she remembers no more until she was picked up and taken on board the *Thingvalla*. She thought at first that the children were also saved, but soon afterward learned they were lost.

Within five minutes of the collision, the *Geiser* sank. Captain Moller recalled,

I jumped on the rail and saw that the vessel was going down. I stayed there until I was swept away by the water. I went down with the vessel, being sucked in by the rushing waters. It seemed to me that I was more than a minute under the water, whirled head over heels, striking objects living and dead. At length I felt I was rising. I did not lose consciousness at all and suddenly shot up to the surface. I at once struck out and got ahold of an oar, which I clung to swimming to support myself for about twenty-five minutes. I could see the lights of the *Thingvalla*, and was surrounded in the water by struggling human beings and floating barrels and boxes.

Wilse, too, was sucked down with the ship, then popped to the surface with the help of his cork-belt life preserver. He wrote, "There was a lot of screaming and moaning, and we saw many heads breaking the surface just to go down, never to come up again. The worst site we saw was SS *Thingvalla* steaming away from us. What use had it been to fight for our lives, when, in a while we would die because of the icy cold water?"

Eventually the *Thingvalla* came around and rescued a handful of survivors. But in all, more than one hundred drowned. The grieving Mrs. Lind was returned to her husband in New York, and the *Times* reporter witnessed a tragic reunion. "Her husband, too, is almost prostrated, and when seen last night the couple did nothing but moan and cry."

Of course, many Norwegians arrived safely in the United States, and by the 1870s, they were flooding the booming timber port of Seattle. "Those Norwegians are as good a class of immigrants as our Territory can possibly be peopled with," gushed Tacoma's *Weekly Pacific Tribune* in 1876, "as they are noted the world over for their industry, economy, honest dealing, and steady habits." They were called squareheads, a nickname whose origin is unclear. Some say that the Plains Indians believed that the Norse settlers were honest and would give them a "square deal." Others say they gained the nickname from being a bit slow upstairs, the opposite of "sharp."

Norwegians settled just northwest of Seattle, along Shilshole Bay and Salmon Bay. Families like the Olesons and the Schillestads staked claims as early as the 1870s. Salmon Bay was a logging and fishing camp, and its first boat builder, T. W. Lake, set up shop in 1872.

Salmon Bay was undesirable real estate because it lacked fresh water for drinking and a deep-water port for big vessels. Only small boats could land, and only at high tide. When in 1887 a local partnership of large landowners decided to disband, they flipped a coin to see who got stuck with the lousy 160 acres along Salmon Bay. The loser was William Rankin Ballard, the former captain of a sternwheel steamship called the *Zepher*. His town, Ballard, was incorporated in 1890, and ten years later had more than 4,500 residents, making it the seventh largest town in the state. The lumber mills thrived—Ballard was king of the cedar shingle.

Norwegian immigrants worked as farmers, loggers, builders, fishermen, and just about everything else. In 1905, Sivert Sagstad arrived and started the Ballard Boat Works on Shilshole Bay. Lars Brekke opened an iron and metal foundry in 1900. Squareheads started Seattle Ship Supply and Nordby Supply Company. A Minnesota-born Norwegian named Albert Julius Hansen—no relation that I know of—arrived in 1890 and later became a Seattle police officer. A. B. Wilse, who survived the sinking of the *Geiser*, moved to Seattle to begin his photography career. Norwegian Lutheran churches appeared as early as 1888. The Leif Erikson Lodge of the Sons of Norway fraternal club was founded in 1903, the biggest such lodge in the country. By 1910 there were seven thousand Norwegians in the Seattle area, and by 1920, five percent of Seattleites were Norse. In 1918, Seattle elected its first Norwegian-American mayor, Ole Hanson (also no relation).

From the start, Ballard had a drinking problem. Like a lot of frontier boomtowns, it was thick with saloons, gambling halls, and whorehouses. Legend says that Ballard had more saloons and more churches per capita than any town in America. In 1904, the mayor ordered the entire town shut down for a day to stop the gambling. It didn't work.

Ballard's main problem was its lack of fresh water. So in 1907 Ballard voted to be annexed by Seattle, and reap the benefits of big-city sanitation. Yet Ballard still lacked a deep water port.

Before the automobile, most of Seattle's commerce and transportation was by water. The city was looking to connect Puget Sound to its inland waters, Lake Union and Lake Washington, perched twenty-two feet above sea level. The solution lay in Ballard.

In 1915, the Army Corps of Engineers descended upon Ballard with steam shovels firing and work crews heaving dirt. Salmon Bay was transformed into a series of locks that would raise or drop ships the necessary twenty-two feet. The shipway was named after Corps Engineer Hiram M. Chittenden, but locals have always used its more common nickname, the Ballard Locks. To the east, a ship canal was dredged from Salmon Bay to Lake Union, and from there to Lake Washington. On July 4, 1917, a small armada led by the *Roosevelt* passed through the locks. Overnight, boats could pass easily between the sound and the lakes, getting their tickets stamped in Ballard along the way.

With the Seattle and Tacoma ports already developed as freight harbors for large craft, Ballard was the ideal port for midsized commercial fishing vessels. In 1930, 200 of the northwest's 300 halibut schooners were berthed in Ballard. The Port of Seattle built a new wharf at Fisherman's Terminal, at the south end of the Ballard Bridge, to accommodate the vessels. The Ballard First Lutheran Church began the tradition of blessing the fleet as they set out each spring. Fishing rivaled the lumber mills as Ballard's top industry, and ever since, Ballard has been the heart of the North Pacific's commercial fleet. To this day, we still take the *Northwestern* through the Ballard locks every time we leave for Alaska or return to Seattle. It's a great feeling to get back to the locks, and your wife and kids waiting at the dock. You know you're home. Without those locks, the fishing industry might have ended up somewhere else, and this saga would be completely different.

Once the locks and wharf were complete, a new industrial zone flourished on Salmon Bay. The dusty streets were paved for automobiles. Sivert Sagstad moved his boat works to the freshwater side of the locks, renaming it Sagstad Shipyards, and went on to build more than three hundred boats. Ballard Oil began fueling ships from a dock in 1937. Ballard survived prohibition and the Depression. World War II got the local factories revving again. Wooden-hulled minesweepers were dispatched for war efforts from Ballard shipyards. With victory, another fishing boom followed. Heitman "Hat" Thompsen, a commercial fishermen from Norway, opened up his marine hydraulics shop in 1946 on Leary Way. Pacific Fishermen built a shipyard in 1948, quickly followed by Western Tugboat Company, and Marine Construction and Design. The waterfront churned out machinery for the maritime industry: Brekke Company Miscellaneous Steel Fabricators, Ballard Sheet Metal Works, Bardahl Manufacturing, Salmon Bay Sand and Gravel, Ballard Pattern and Brass Foundry.

The surrounding neighborhood, where Ballard Avenue, Leary Avenue, and Shilshole Avenue run at a 45-degree angle to the city grid, became the heart of the Norse community. The brick lanes, iron curbs, and brick storefronts crowding the streets even looked a bit like Scandinavia, especially with the sea smells on the wind, the ships on the water, and the dancing sounds of Norwegian and Swedish spoken in nearly every shop. The second generation created their own hybrid language, Ballard Norwegian, a mix of English and Scandinavian words. The stores were practical, workingman stuff: Olsen's Furniture Company, Eidem's Upholstery, Ballard Hardware, and Nielsen Brothers Carpets, which would tack carpet samples on Ballard Avenue telephone poles to attract attention. The JCPenney on Market Street, which opened in 1924, sold more black wool union suits than any other Penney's in the country. The neighborhood gained the nickname Snoose Junction, for all the *snus* tobacco that the Scandinavians chewed.

The squareheads were notoriously frugal. Many who grew up in Ballard with immigrant parents remember Dad ordering them to spend the afternoons straightening nails that would someday be reused. Even when times were good, the fishermen would walk up Ballard Avenue jingling change in their pockets, reeking of fish and diesel. "Bad year," they'd complain. Then they'd drive off in their Cadillacs.

This was the world Dad entered in 1958, twenty years old and hungry, his first time out of the provinces of Scandinavia.

Sverre and his friend Tormod Kristensen walked the docks. The two lifelong friends had grown up together in Karmoy, gone to school together, and fished together. Tormod had relatives of his own who'd offered him a place to live.

They started on the Ballard locks, then walked east to the boat-yards. From there it was a hike to 15th Avenue where they crossed the ship canal on the Ballard Bridge, and walked the wharf at Fishermen's Terminal. They spoke little English, but many fishermen were squareheads, anyway. Even then it was hard to get a job. A lot of the old-timers had been around for ever, and were very clannish. Some of those crusty old squareheads never got off those boats. The money was good, but not so good that skippers were building more boats.

You had to kill people for a job back then. Even with an uncle in the fleet, Sverre couldn't find a job. So the two young sailors beat the pavement on Shilshole Avenue, approaching captains and asking for work. My dad used to tell me this story when I was a kid.

"We should go buy some bread so we have something to eat," Tormod said. "We're running out of money."

"So what?" said Sverre. "Let's go up to the corner and buy a bottle of Thunderbird wine for us. We'll figure it out later."

It was just after noon. Tormod shrugged and said why not. So they climbed up Market Street and bought the bottle, and as they strolled back down toward the shipping canal, they ran into the skipper of a trawler called the *Western Flyer*. Two men had quit, and he needed a couple of hands.

"I'm willing to work," said Sverre, and he and Tormod got the jobs.

Years later, Dad used to laugh and say, "If it wasn't for that bottle of

Thunderbird I would have never had my start." (When I asked Tormod about this story, he gave his typical long silence, and said, "I don't remember it quite like that." It doesn't matter, though. Sagas are passed down from father to son, and if my dad embellished a bit, I guess I just add it to his story.)

The captain of the *Western Flyer* was Dan Luketa, and he became Sverre's mentor and hero. An American of Slovenian heritage, Luketa was known in the fleet as a tough son of a gun. His brother Paul had been lost at sea years earlier. Dan Luketa suffered no fools, and he tolerated no slack. "He was a good man, a hard driver," says Tormod. "You always made money with him." In those days all they did was shake hands, none of these goddamn lawyers, but you may as well have signed a contract. His word was good.

Luketa had dark hair, square shoulders, and chewed a cigar. He wore thick, heavy glasses, and sometimes he brought his pretty wife Maxine along on the boat to help him see where he was going. Even with poor vision he was one of the best skippers in the fleet—a highliner. He owned three boats, the *Sunbeam,* the *Paul L,* and the *Western Flyer.* Luketa would go out drinking with the deckhands, dancing with his wife, but if you didn't show up on time the next morning, you were fired, just like that. You didn't do him no wrong. A lot of guys didn't like him, thought he was too stern, but he was straightforward and fair. Sverre was drawn to him like a son to a father.

Sverre started working on the *Western Flyer,* and Tormod on the *Paul L,* dragging the coast for cod, perch, dover sole, and flounder. They longlined for halibut, and fished for some salmon, too. In those days, American fishermen were allowed to fish Canadian waters, and vice versa, so they would drag off the coast of British Columbia, up to Hecate Strait and off Vancouver Island.

Dad got his first job in America aboard Dan Luketa's *Western Flyer.* *(Courtesy of the Hansen Family)*

The *Western Flyer* already had a storied past. It was the vessel that John Steinbeck and Doc Ricketts chartered for their 1940 voyage up the Sea of Cortez to catalog marine species. The seventy-five-foot wooden seiner was part of the sardine fleet out of Monterey, California. Steinbeck and crew sailed from Monterey along the coast to Cabo San Lucas, then all the way up to the Gulf of California. Steinbeck's book about the two-month journey, *The Log from the Sea of Cortez,* is still in print today.

Work on deck the *Western Flyer* was hard. When they were bottom-trawling for dover sole—or "slimeys"—they'd drop the trawl net three hundred fathoms, which required more than three-quarters of a mile of steel cable. The men were so exhausted they'd fall asleep trying to lead that much cable into the water by hand. Later it would take forty-five minutes to haul it all onboard.

Nearly every time the net came up, it was torn to ribbons. Luketa would take one look at the heap of tangled mesh and start barking out orders—he knew exactly how to put it back together. He drove the crew hard. More important, there were always plenty of fish in the bag. Luketa knew where to drag his net. He was known for working twenty hours without a break. He'd bring one boat in from a week in Canada, then jump on his other boat at Fisherman's Terminal, and head back out without a night in town. Nobody worked as hard as Luketa, but if you wanted a job on his boat, you'd better try.

Luketa was also an innovator, a mechanical genius. If some part of the boat didn't work as well as it could, he'd sit there for hours, staring, trying to figure something out. Then he'd get an idea, build it, and try it. Some things worked, some didn't. He invented and patented a few fishing innovations, including a door that held a dragging net open, which came to be called the Luketa Doors.

Back then the fishing in Seattle was pretty rough. Magne Nes, who had fished on the East Coast, worked with Luketa and Sverre in the late fifties. "In New Bedford, we were pretty much civilized," he says. "We bled the fish, we gutted them, then we went to the market. If there was just a few marks, then the fish were half price. So we treated the fish nice on the inside after we gutted them. Here, it was just garbage. We didn't treat the fish very good. We had a union in New Bedford, but here it was the Wild West. There was no union, nothing. I remember we made a big haul of red fish—must be fifty thousand pounds of fish. We called on the radio to tell the buyer what we had. But the fish were not big enough, 'No we don't want to buy it.' So we had to dump them. It was a lot of fish. The whole ocean was just red with fish. And we had to find some bigger ones."

If Seattle was the Wild West, Sverre was starting to like it. He wrote letters home bragging to Karl and their father about all the money he was making. He learned to play guitar and fell in love with American music, especially the country-western and rockabilly sounds of Johnny Cash, Hank Snow, and George Jones. His favorite was Johnny Horton, the Texas twanger whose "saga songs" raced up the charts in 1960. First came "The Battle of New Orleans" about Andrew Jackson in the War of 1812, and then "Sink the *Bismark*," a kitschy tale of World War II heroics and revenge that a kid raised under Nazi occupation, who idolized Shetlands Larsen and Winston Churchill, just couldn't resist:

> *We'll find the German battleship that's makin' such a fuss*
> *We gotta sink the* Bismark *'cause the world depends on us*
> *Yeah hit the decks a runnin' boys and spin those guns around*
> *When we find the* Bismark *we gotta cut her down*

That fall Johnny Horton was killed in a car wreck, and days later his latest—and last—song hit Number One on the charts. "North to Alaska" was a catchy ditty that Sverre would count as his favorite for years to come. It may have even inspired his next big adventure:

> *Big Sam left Seattle in the year of ninety-two,*
> *With George Pratt, his partner, and brother, Billy, too.*
> *They crossed the Yukon River and found the bonanza gold*
> *Below that old White Mountain just a little south-east of Nome.*

Dad *(right)*, while serving in the U.S. Army, 1962. *(Courtesy of the Hansen Family)*

The next year, Sverre sponsored his brother to join him on the *Western Flyer.* Sverre was envisioning a future in America: good work, good money, good music, and now even family, too. Yet something happened that would test his affection for his new country. Sverre got a notice from Selective Service: He'd been drafted. Even though he wasn't a citizen, as a legal worker he was fair game. "Just like Elvis Presley!" he snorted to his buddies. "Plucked in my prime!"

Sverre wrestled with the decision. He could avoid the draft by going back to Norway, but then he might never be allowed back. Yet part of him was drawn to the adventure of the military—just like his hero, the naval captain Shetland's Larsen. He asked Luketa for advice.

"It's your duty," Luketa said, "if you want to be an American."

Luketa pointed out that by enlisting, he could become a legal citizen sooner, thus allowing him to be a skipper.

"If you do the right thing, and enlist," Luketa said, "I promise you'll have your job when you return."

Sverre was inducted. After basic training he was shipped to Berlin, where he would be stationed for two years. After a childhood under German rule, it had to give him some satisfaction to pay his first visit to that hated country as a uniformed soldier—in the occupying army.

In the fall of 1987 I was in Seattle getting ready to head back up to Dutch Harbor for king crab season. Tormod was going to be captain, and the crew was Brad Parker, Pete Evanson, and Mark Peterson. We moved the boat through the locks and tied up at Shilshole Marina, planning to leave first thing the next morning. Shilshole is a yacht harbor for pleasure craft. Fishermen don't usually use it. On the dock looking out to Puget Sound stood the statue of Leif Erikson in his iron helmet, sword, and battle axe, with a cross hanging from his neck. It's a towering thing, at least twelve feet tall. Out on the point at Golden Gardens, teens were burning bonfires and drinking strawberry wine. Up above the marina, the houses seemed to look down on us through the thick trees. We would all sleep on the boat that night and motor north in the morning.

In those days I was feeling pretty confident—maybe even invincible. I was making good money, and had finally bought that Corvette I always wanted. It was a midnight blue 1982. I would race that thing around Edmonds at ninety miles per hour. Within six months of buying it, my buddy Leif challenged me one night in his Trans Am. As I ripped around a corner I lost control, jumped the embankment, knocked down a chain-link fence, and soared down a twenty-foot drop. The car spun 180 degrees and landed backward, flat as a pancake, with all four wheels snapped at the axle. Leif looked down from above and said, "You lost." I climbed out of the driver's seat unscathed—we thought the whole thing was funny as hell. With all the dangerous things I did at work and in cars, I never thought much about it catching up with me.

That autumn night down in Shilshole Marina, a crabber/trawler called the *Nordfjord* was tied up next to us. The boat was 127 feet long and just eight years old. It was owned by Agust Gudjonsson, an Icelander who came over to Seattle about the same time as my dad. His son, twenty-eight-year-old Gudjonroy, whom everyone called Roy, was the skipper. Roy and his wife were expecting their first baby in a few months. Our crew and their crew had been hanging out for the last few days in Ballard, rigging and loading the boats, and shopping for groceries. Like our crew, theirs was also young. The oldest onboard was an Icelandic guy who was forty, but the others were in their twenties or early thirties. We ran together through the locks. Our boat was tied up at the A-dock, and they tied up outside of us, so they had to cross our boat getting on and off. The *Nordfjord* was headed up to Alaska to fish. Our two ships would motor up together on the seven-day trip to Unimak Pass and the Bering Sea.

The *Nordfjord* had been in dry dock for regular servicing and was in great shape. It was equipped for the worst of the Bering Sea, with auxiliary power sources, backup radios, fire sensor alarms in the engine rooms, and firefighting equipment. It was stocked with life rafts and survival suits for the entire crew, as well as two emergency position-indicating radio beacons, or EPIRBs. These devices were to be manually activated if the vessel was sinking and lost radio contact. EPIRBs send a satellite transmission that is relayed to the Coast Guard, indicating time and location of the distress call, so that a search can begin.

That night we were all drinking in Charlie's, the bar in the marina. I remember it well because across from the bar, in the restaurant, my parents were having dinner with Tormod. The rest of us young guys were throwing back drinks. I had turned twenty-one only a few months before, and I'd never been in a bar with my mom watching me. It was weird. She kept looking over at me, like she was counting my drinks and thinking: *Shouldn't he be getting to bed soon?* I just ignored her.

So there we were, the deckhands, shooting the shit like we always did, talking about how much money we were going to make and what we'd do with it when we got home. I liked all those guys. We were having a good time, swapping stories and probably telling some lies. We closed the bar down and crawled onboard and slept on the boats.

The next morning we fired up the engines. The *Nordfjord* was running great, but the *Northwestern* had some mechanical problems. Tormod and Pete went down to the engine room to look over the engine, but they couldn't fix it. So they called a mechanic. This was going to take a while. Meanwhile, Roy and his crew were getting restless. We'd already spent a whole night tied up in Shilshole.

"We'll see you up there," Roy called. The guys untied and steamed north up the sound. We figured we'd see them soon in Dutch. It took all day, but we got our boat fixed. About twelve hours later, we radioed the *Nordfjord* and told them we were on our way.

A few nights later the *Nordfjord* hit some weather north of Vancouver Island, but the 20-foot seas then subsided. At 9:30 P.M. on September 18, Roy radioed his brother and said that the forecast was calm and they were heading across the Gulf. Just hours later, at two in the morning, the Coast Guard station in Kodiak picked up a frantic message.

"Mayday, Mayday, Mayday," called the out-of-breath skipper. "This is the fishing vessel *Nordfjord*. Mayday, Mayday ... Mayday. Mayday. Over."

That was all. The Coast Guard couldn't get another response from the ailing boat. They launched a search. The next day two Coast Guard and a Canadian aircraft crisscrossed the ocean west of Vancouver Island, near the *Nordfjord*'s last known position. The searchers spotted some oil drums, mattresses, and a wooden ladder, but could not confirm that they belonged to the missing ship. Two days later a fisherman found a float from the *Nordfjord,* but nothing more. The boat had sunk so quickly the crew hadn't had time to activate the EPIRBs. The $1.5 million state-of-the-art boat with its five crewmen had simply vanished.

It bothered me for a long time. I had just been with these guys. They had become friends. As far as I know we were the last people to see them alive. I kept thinking that if we'd left together we would have been right there. We should have been side by side crossing the gulf. We could have helped them. The loss of the *Nordfjord* hurt. A few years ago the producers asked me about it on the show, and I got choked up just talking about it. It brought back a lot of stuff I hadn't thought about in a long time. If our engine hadn't died we would have been there for them.

Like my father, I've never gone to church regularly; it's nearly impossible when you're at sea most of the year. In spite of that, I'm a believer. The power of the ocean makes me realize what a tiny place I occupy in the universe, and that despite all our best efforts, we have little control over our destiny. The work I do has forced me to contemplate death—to wonder why one man lives and another man dies. The answer to these questions is simply beyond the grasp of any mortal. All I can do is pray for my safety and pray for the safety of my family and crew.

4

NORTH TO ALASKA

D renched to the skin, his trousers stiff with frost, Captain Sverre heaved the rubber hose to Krist. "Now!" Sverre cried. Krist stuffed it in the engine hatch, dousing the fire below. It wasn't working but there was no other plan. Then Sverre heard a sound—like shattering glass—coming from the wheelhouse. The windows were blowing! He dropped to a knee and covered his head, but no shards of glass tore open his flesh. Sverre looked up. As dawn spread across the sea, the dark outlines of the ship's silhouette had faded, and for the first time that day he could see clearly. The shattering sound was not glass—it was icicles dropping from the eaves and exploding on deck. Why now? Was it just the wind? Maybe daybreak was heating the air imperceptibly. That was wishful thinking. As it turned out, the icicles were suddenly releasing their grip because the fire below deck was heating the boat. Conduction. Another icicle leapt from its hold and smashed on the deck.

Sverre and his crew fought the fire to something like a draw, which is to say the boat didn't explode, and the flames didn't come above-deck. However, the captain didn't feel like he was winning this fight. Each time they opened a new air vent, the flames leapt up. The value of flooding the vent with water was offset by the addition of oxygen, which simply fueled the hungry fire.

Still Captain Sverre did not radio a Mayday. It wasn't that he didn't know the danger—he just refused to lose the boat. The minutes required to climb the wheelhouse and call for help were minutes he could be fighting the fire. They labored on, soaking wet in their cotton pants and shirts, simultaneously freezing from the arctic wind and sweating in a fit of fatigue and adrenaline. An oily black ribbon of smoke rose off the boat and swirled in the gusts of wind. Belowdecks, those thousands of gallons of diesel fuel brewed in their tanks. So far the fire hadn't reached them. So far.

Maybe they needed to fight the fire more directly—to haul the hose down the stairs into the engine room. It was almost suicidal, but it seemed the only choice. Captain Sverre ordered Magne and Krist into the galley to see if they could access the engine room. The men rushed in, and almost immediately turned back. The galley was flooded now with thick smoke and the men were coughing. They couldn't get inside. Not only were they blocked from the engine room—they'd also lost their last chance to salvage warm clothes, raingear, and food from the galley.

Beneath the thin soles of the captain's slippers, the wooden planks were hot. The raging fire was spreading. Since he couldn't actually see it, he could only guess what had burned, and where. One unlucky step could crash through a board smoldering on the underside. If a man fell through the deck into the furnace, it would be impossible to rescue him.

The situation was desperate. A few more minutes and the wheelhouse would fill with smoke. Then the captain wouldn't be able to get off a Mayday. It was now or never.

"Take the hose," he called to Leif.

Sverre ran to the ladder and climbed to the upper deck and pushed into the wheelhouse. Smoke gushed from the galley below, but he could still see, and still breathe. He rushed to the captain's chair and snatched the radio from the wall.

"Mayday! Mayday," he cried. "This is the F/V *Foremost!* Mayday." There was no response, not even the beep and squelch of his own transmission. Sverre's eyes followed the coiled cord of the handset to the radio base mounted overhead. No lights. No nothing. He flipped the power switch. It was dead. He scanned the bank of electronics. Everything was dead. No radio. No power. Captain Sverre and his men were all alone and in big trouble. The arctic gale whipped the flames on their wooden boat as it pitched madly in the 30-foot seas.

As I've said, this saga is about brothers. Our story, and our dad's story, is closely tied to that of his brother Karl. Just as Dad came to America looking for a better future, so did Uncle Karl. In 1961, just before being inducted into the army, Sverre had sponsored his brother's immigration. On July 7, Karl Johan Hansen flew from Copenhagen to Reykjavik to Los Angeles and finally to Seattle. Sverre was at sea, and Uncle Jorgen was supposed to meet Karl at the airport, but Jorgen didn't show. So Karl hopped in a taxi and handed over a scrap of paper with the address scrawled on it. Karl spoke even less English than Sverre.

When he got to the house the kids were playing on the lawn. As it turned out, Jorgen had gotten stuck in traffic. When Jorgen returned, he made up for it by taking Karl on a driving tour of Seattle. He took the kid up and down hills and across bridges. Karl had no idea where he was. Then Jorgen drove them down by the waterfront, to a strip of seedy bars on a crowded brick street: the Smoke Shop, Ballard Tavern, Vasa, and Malmen's.

"This is Ballard Avenue," said Jorgen. "It's a good place to stay away from."

"Oh," said Karl, watching the drunks stumble down the sidewalks. "Why's that?"

"Well, it's where all the fishermen go and drink."

Karl was only nineteen years old, not old enough to legally drink in Washington State. Uncle Jorgen drove him home. Karl told his aunt and uncle he was going out for a stroll. On Third Avenue he hailed a cab.

"Ballard Avenue," he told the driver.

Months before, Karl had been home in Karmoy struggling to make ends meet. He had fished herring with his father two seasons in Iceland. The first year they went purse seining, the second they went drift netting. Sigurd had become a fishmaster like his own father, but the fishing was mediocre and they didn't make much money. Karl judged it to be miserable work. He then joined the merchant marines as a galley cook, and when he came home at age eighteen he was hired to cook on a dragger called the *Havbell*.

The *Havbell*, which translates to "Ocean Belle," was moored in the Karmoy village of Skudeneshavn. One nice summer day Karl took a bus there. The owners had just bought the boat cheap, and it needed work. Before they could launch, the crew had to replace the engine, a job that would last a few days. Most of the men wanted to sleep onshore, but since there were nice benches in the galley, Karl volunteered to sleep on the boat.

"Are you sure you want to do this?"

It was a beautiful sunny day. Karl knew a lot of girls in the village.

"Yeah, why the hell not?"

"OK. You're sure?"

So he slept on the boat. In the morning the men returned.

"Have you heard anything?" they demanded.

"No."

"Have you seen anything?"

"No."

Then they told him the truth. The boat was rumored to be haunted. They'd been using Karl as a test to see if he was visited by the ghost. Later that year when they sailed in bad weather to Denmark and Sweden to pick up their nets, one man claimed he'd seen somebody on the boat. The man swore it until the day he died. Karl figured it was just a dream brought on by the storm.

But Christmas Eve, 1960, the *Havbell*'s bad luck caught up. The ship with its six crewmen was rounding a point with a lighthouse when it ran aground. It was cold as hell, dark, wind blowing, swells rising, and the boat lay down on the rocks. The skipper figured the only way to save the crew was to get help from the lighthouse, but the shore was too far to reach by swimming. The boat lay against a big rock. If they climbed out on that, maybe they could make it to shore. The skipper's brother volunteered. He hopped the rail and scampered to the rock with a line. Then two men followed him. Those on the ship shivered in the bitter cold, and strained to see what had become of their crewmen. "It was dark like the grave," Karl remembers. He was the youngest man on the boat, and with one foot over the rail, the boat on its side, he was about to abandon ship for the rock. Then he heard hollering from the darkness: "We're freezing to death!"

One of the older crewmen motioned to Karl to get back onboard.

"Don't go yet," he said. "Take it easy."

Just then a wave crashed over the men on the rock, soaking them to the skin. Now they really were freezing to death, and knew they had precious little time to get to shore. The skipper's brother leapt into the black water and swam. The men waited in fear and looked to the shore for some sign of life. Then they heard a holler. He had made it. Quickly, the second man jumped in, and washed up safely on shore. The third man was older and less fit and he couldn't commit. He sat perched on the rock. By now the skipper's brother was scrambling up the cliffs to the lighthouse. "He was like a cat, that guy," says Karl. The lighthouse keepers couldn't figure out how the hell he'd arrived at their doorstep in the darkness and the storm.

Meanwhile, the men kept yelling to each other in the darkness. Then, all of a sudden, no shouts were heard from the last guy on the rock. He was gone.

The *Havbell* began to break up on the rocks. Waves pounded across the exposed hull. *I'd better get in the life raft,* Karl thought. By then they had inflated the raft and tied it to the bow. Before he could reach the raft, a swell washed him and an empty wooden fish box across the boat. Karl grabbed hold of the mast and clung for dear life as the wall of water rushed out to sea, and someone else—he wasn't sure who—clung to him. When the swell passed, Karl regained his footing and ran toward the raft, but it was gone, ripped loose by the wave. Now there were few options.

Suddenly the men on shore began to holler. They had gotten help from the lighthouse. One of them shot a thin strand from a gun. The line landed on the foundering ship and Karl grabbed it. A thick line was attached to the thin strand, and hauled aboard. Then they attached a seat to the thick line.

"You go first," the men told Karl.

He fastened himself in the rescue chair and gave a shout. The men on shore began pulling, dragging him through the icy water. He coughed and kicked, but finally reached the shore. Then they pulled the final two men ashore. A few hours later, all that remained of the boat was its hull. The pilot house was gone. The body of the drowned man surfaced two months later, miles to the north. As for the life raft that Karl was moments from boarding—it was never retrieved. Karl felt luckier than hell to still be alive.

Needless to say, he needed a new job. Soon after, Sverre came home to visit. He told of the good life in Seattle, and Karl needed no more convincing. He flew to America and took a job with his brother, working on the *Western Flyer.* That's how he ended up on Ballard Avenue.

Karl liked the place. Soon, a bunch of Karmoy boys were hanging out down there. In addition to Sverre and Karl, there was John Jakobsen, Magne Berg, Magne Nes, Krist Leknes, Pete Haugen, Leif Hagen, Arnie Haugen, Sigmund Andreasson, Gunnleiv Loklingholm, Borge Mannes, Tormod Kristensen, to name just a few. The Karmoy boys were strong, lean, hard drinkers and hard smokers. They wore Levis and white T-shirts on the docks, but when they went out on the town they had style: dark suits, white shirts, and skinny black ties, their blond hair combed back into ducktails. They looked like young Marlon Brandos, but for the occasional Norwegian sweater hand-knitted with snowflakes and reindeer.

"These Karmoy people, they're a special breed," says John Sjong, a highliner captain who hails from a different part of Norway. "Good people and good workers. I had Karmoy boys with me the whole time—sometimes I was the only one from the outside. You know, they can be a little obnoxious when they get drinking…"

In those days there were seventeen taverns on Ballard Avenue and three cocktail lounges. A night on the town usually began at the Ballard Tavern, a smoky dive owned by an old Norwegian named Ingmar Boe, who everyone called Inky. He'd been the star quarterback at the

University of Oregon, and then a pro player for the Seattle Bombers, until a broken leg ended his career and forced him to walk with a limp.

Fridays and Saturdays Inky held a dance at the tavern. By law, he had a cop on the premises. He hired Ballard beat cops to work the door in plainclothes. Like the rest of the Ballard police force, the doorman was lenient with fishermen. Everyone was on a first-name basis. Karl wasn't old enough to drink, but that didn't matter. "Go all the way to the back so no one sees you," the cop told him. A lot of the females who came to the dances were working girls from the Boeing plant who had a few drinks and danced, then went home.

On Sundays when the bar was closed, the boys would volunteer to "clean" the joint, which meant taking free pulls off the taps when the old man turned his head. Inky would lend the boys money, or cash their checks. "But only if you spend the money here!" he warned them. Sverre and Karl gave their word. The Tavern only served beer, so after giving Inky the impression they were done for the night, they'd slip down the block to Vasa for some booze.

"You come in here and borrow money," complained Inky, "then you go over there and drink it away!"

"Inky was like a father to us," remembers Magne Nes. On one occasion when Sverre couldn't make it home, Inky took him to his house, moved his daughter to the couch, and let Sverre sleep in the girl's bed. He awoke surrounded by dolls, wondering where the hell he was.

Another night, after two in the morning, Inky was awakened by a phone call. It was Sverre.

"I need to borrow money," he said.

Inky asked for what.

"Bail," Sverre said. "This is my one phone call."

He explained that he had been arrested for fighting and hauled downtown. Bail was $75.

"Oh, and while I have you on the phone, there's someone else who needs to talk to you."

He passed the phone to Karl, and then three other Karmoy boys. They'd all been thrown in the tank. Inky bailed them out.

Liquor laws allowed only restaurants to pour the hard stuff. So Malmen's Fine Food served dinner in the back and cocktails in the front, while the Smoke Shop had its "Amber Room" and Vasa Sea Grill had its "Patio Room." Vasa was just three doors down from the Ballard Tavern, so the boys hustled in there. It was named after an infamous Swedish warship built by King Gustavus Adolphus in the seventeenth century. The wooden boat was 230 feet long and 172 feet tall, weighing 1,200 tons and capable of carrying 450 men. The king rushed the construction because he wanted to send it off to the Thirty Years War. After three years of building, she launched on her maiden voyage with royal fanfare. The *Vasa* had sailed less than one nautical mile when a mild breeze picked up, and the top-heavy boat with no ballast foundered and sank to the bottom. So much for the Swedish armada.

The Patio Room was smoky and raucous, crowded with old sailors and fishermen, most of them Scandinavian. Many rented rooms in the run-down boardinghouses above the storefronts—the Princess Hotel, the Sunset, and the Starlight. Back then halibut season lasted six months, and so for the other six months those fishermen would just hole up in their hotels and spend their money at the bars. One old-timer just deposited his paycheck at Malmen's, and drank from his tab until it was gone. The crowd was almost all men—in those days women were not allowed to sit at the bar or carry a drink from table to table, a holdover from the antiprostitution laws from decades before. Out the back window of Vasa's Patio Room a man could see the Ballard locks and the Pacific Fisherman shipyard, the Scandinavian crosses fluttering in the wind, and Queen Anne hill rising out of Salmon Bay. Between the lounges and the tavern, the boys had little reason to leave the block, let alone the neighborhood.

As last call approached, they'd hike two blocks down to Malmen's, another Swedish-owned dive, where they got martinis for seventy cents. The beer and ale selections were Western, Eastern, and Lowenbrau imported. For eighty-five cents the menu offered Aalborg, a Danish brand of aquavit, the traditional Scandinavian liquor flavored with caraway seeds, which was to be "Enjoyed Before Meals." For an extra nickel, you could get the Swedish liquor, Carlshamns Flagg Punch—to be "Enjoyed After Meals." Liquor had been served in this location continuously since 1893, a feat memorialized in the storyboard of cigarette burns pocked in the magnificent handcarved wooden bar.

Though the club closed at two, the restaurant was open all night, and offered Sverre, Karl, and the boys its Famous Salted Boiled Black Cod, with potatoes and drawn butter for $1.75. They shoveled the fish into their mouths in booths beneath a mural that spanned an entire wall, a pastoral landscape of blonde maidens blowing horns in an alpine meadow beneath jagged snow peaks. One such fräulein leaned against a fir tree and gazed across the lake at a farmer whipping his buggy horse through the fields, while a white-headed shepherd in suspenders, slumped in resignation, and watched the whole scene unfold.

Karl worked on the *Louie G* with Jackie Ray, a tough old sailor with a hook for a hand. Jackie was from the East Coast, and had lost his hand when he got it twisted in a net-hauling drum. The drum ripped it right off. Ray liked to sit in the Smoke Shop and challenge men to arm wrestle with his good arm. He usually won. If he lost, or if he simply didn't like you, he'd come after you with the hook. He could tear your nose off with that thing.

One day after an arm wrestle, Karl said to him, "Why don't you put up the other one?" He knew that Jackie would come up swinging after a taunt like that, so he grabbed hold of the hook on the table before Jackie could make a move. Jackie laughed with hatred in eyes. He was dangerous. Magne Nes remembers working as a greenhorn on Jackie's boat, the *Havana,* and watching Jackie and another deckhand go after each other in a bloody fight. Jackie wielded his claw and the sailor used a gaffing hook. You never knew what would happen with that guy.

Years later, Jackie Ray was skipper of the *Western Flyer* when it had been converted for crab. That was back before they had the hydraulic pot launcher. Ray would help the deckhands push the pots over the rail and into the water. One day, his hook got caught in the mesh and he was pulled overboard. The men hurried to feed the line in the block and pull the pot. All that surfaced was the hook and the harness, rattling in steel bars. He must have had the strength to free himself from his hook, but not enough to swim to the surface. They never found his body. Some wonder if those deckhands hadn't had enough of Jackie Ray and given him a little help to the bottom.

While Sverre was away, tragedy struck. Uncle Jorgen was out on a trawler called the *Guide.* The steel cable used to haul nets was rotten, but instead of replacing it, the owner spliced together the existing strands. One day out in Puget Sound they snagged something heavy in the net, probably a big rock. They hauled the net to the surface and attached it to the block—the one with the rotten cable. The machinery strained and whined. The cable hummed. Suddenly it snapped. The backlash in the steel line blindsided Jorgen with a blow to the head. He fell, probably unconscious, tangled in the net. Free from the block, the net raced to the bottom of the sea, with Jorgen attached. His body was never found.

Jorgen's widow and children returned to Karmoy. Now Karl was alone in Seattle, awaiting the return of his brother.

Maybe the best part of being on a crab boat is the guys you end up working with. My family has very high standards for who we hire. We don't just take guys off the dock. I've done that a couple of times, but it didn't work out. Once, I hired a bartender who pestered me for two years for a job. After two days at sea, we had to drop him off at port because he hated it so much. I don't blame him. It's not for everybody.

We trust these guys with our lives—literally—so we hire the best. When it's time to hire, I already have someone in mind. We have to know about a guy, and he has to come highly recommended, before he gets a slot on deck. Our current crew, in addition to my brothers and me, are three of the best deckhands in the business: Matt Bradley, Nick Mavar, Jr., and Jake Anderson.

Matt Bradley grew up in the neighborhood. He and Edgar have been friends since junior high. Back when Edgar didn't want to be a fisherman, Matt got him his first job—working in a restaurant. Then Edgar returned the favor and the Old Man hired Matt to rig gear and run shots in the summer when we were rigging our pots. Most of the people in our school had plenty of money. My dad was doing well, so we weren't suffering. Matt came from the other side of town and had a rough family life. "I lived in a two-bedroom piece of shit," he says. "It used to be a cabin." Like Edgar, Matt didn't have much use for high school, and never finished. My dad always had a soft spot for an under-dog, and he took a liking to Matt. I helped get Matt hired on another Norwegian crab boat, and then we hired him on the *Northwestern*. He's been with us more than fifteen years, one of the most steadfast and loyal deckhands in the boat's history.

Of course there have been rough spots, fistfights and everything else. As he's talked about on the show, Matt has battled drug addiction for most of his life. For years he was making great money with us, but within a few weeks of payday he'd blow it all on drugs, girls, and hotel rooms. While the rest of the crew was buying their second or third home, Matt was living in his van. I tried to work out a better system for him. I offered to pay him just a couple hundred per week, and keep the rest in an account for him, but Matt wanted the twenty grand in a lump sum. The lump sum would always disappear real quick. Then he'd miss the flight up to Alaska and we'd have to wait two days for him to catch up. Once he missed the whole season. Finally I couldn't do it anymore. He came to me for an advance.

"You look like shit," I said.

"I'm fine."

"What are you going to do different this time?" I said.

"I'm fine."

"You're gonna die and I'm gonna be responsible," I said. "It's like I'm signing your death warrant on these checks."

So I fired him—but found him work on another boat. Eighteen months later he called me. He'd been clean for eight months. He wanted his job back on the *Northwestern,* and I was happy to have him. He knew the boat so well, and worked so hard—he made the ship run better. "The guys who were filling my spot couldn't fill my boots," he says. He's been with us ever since, and just celebrated five years in recovery. Since he's talked about recovery on the show, and wears a twelve-step program symbol on his jacket, fans have begun contacting him through the Internet, asking for help with their own addictions. Just recently a guy he'd been talking to online called on the phone. He'd been clean five days. Matt agreed to meet him at a meeting in the neighborhood. "Someone did it for me," he says, "so I want to help someone else out."

Our oldest crewman is Nick Mavar, Jr., who comes from a proud family of fishermen. He and I are the same age, and like me, his parents are immigrants, his mom from Ireland and dad from Croatia. Nick grew up in Anacortes, Washington, in a community of established fishermen, many of them Croatian. His father owned a salmon boat in Bristol Bay, and beginning at age eleven, Nick was up there working summers. He and his brothers were the crew.

In 1983, Anacortes lost fourteen men when two local boats, the *Americus* and the *Altair,* sank near Dutch Harbor. It was one of the worst disasters in crab fishing history, and Nick's closest high school friends were among those lost. Even so, Nick took a job on a crab boat that same year. "That's just fishing," he says. "There's always some disaster. That's the risk you take." As it turned out, Nick didn't take to crab fishing and after a season decided to focus on salmon. He bought his own boat in 1988. A few years later, he gave crab fishing another try on a highline boat called the *Lady Anne,* He's stuck with it ever since. Seven years ago, we had a deckhand quit before cod season, and Matt Bradley suggested Nick as a last-minute fill-in. I hired him, and he's been with us on the *Northwestern* ever since. He is as solid and dependable as they come.

Nick's father and brothers still fish up in Bristol Bay. He has a son and a daughter, but says he's not sure his son wants to follow into the family business. Time will tell.

In Nick's family we found another great fishermen. A few years ago when we were looking for a greenhorn, Nick suggested his brother's nephew, Jake Anderson. Like his uncles, Jake comes from Anacortes. He was in first grade when the *Americus* and the *Altair* sank. He remembers the kids from his class who lost their uncles. The whole town was in mourning. Jake still wanted to work on boats, though. "In Anacortes you're either a fishermen or you work in the refinery," he says. Jake began fishing for salmon in Bristol Bay when he was in high school.

When he graduated in 1999, Jake had interests other than fishing. He was a semiprofessional skateboarder, landing sponsorships and appearing in magazines. His specialty was launching. From his days on a snowboard he was used to hucking thirty-foot cliffs. In 1999 he set a record when he flew over twenty-one steps on his skateboard. Right after that, he fell while jumping off a mere nine steps, and shattered his ankle. It was a slow recovery. By the time he healed, his friends had surpassed him, and he was frustrated. "Mentally I let it get to me," he says.

That's when he got more serious about fishing. After five years of gillnetting in Bristol Bay, he landed a job on a cod pot boat in Kodiak, but he wasn't making money. The owner was taking advantage of his inexperience and not paying him fairly. His Uncle Brian told him to quit, and he did. Later he found work on a crab boat, and that's when Edgar offered him a slot on the *Northwestern*. At first he didn't want it. He thought he'd rather be on a dragger. Also, by then, *Deadliest Catch* had already begun. "I figured they were going to beat me up on purpose on TV," he said.

Of course, we didn't. We liked him. He was a good worker. He's got a great attitude. We invited him back for opie season—and he's been with us for three years. We still call him the greenhorn, or "The Kid," but he's a full-share deckhand. He's like family now. Last winter Jake got his AB license—Able Bodied Seaman—and is planning to get his mate's license. His goal is to someday be skipper of a boat. I think he has everything it takes to do it.

So that's our crew: Edgar, Norman, Matt, Nick, and Jake. I talk a lot about what happens at sea, and it won't make sense unless you understand how a crab boat works. I'll explain that now to get it out of the way.

When we set out from port in the *Northwestern,* we have as many as two hundred crab pots stacked on the deck. *Pots* is the word they've been called for ages, from when they were small round baskets that a fisherman could haul himself. These days, crab pots are more like cages. They are box-shaped, seven feet square and three feet deep, and weigh between 700 and 850 pounds empty. A frame of steel bars, welded at the corners, is wrapped with nylon mesh. A hole in the mesh wall allows the crabs to enter, but not escape.

The first step is to set the gear, which basically means dropping the pots overboard. On our boat, the deckhands sometimes alternate jobs. To make it easier to understand, I'll explain it as if the same guy always does the same job. So to start, Nick Mavar scales the stack of pots, which is three stories tall. When you add in the height of the deck, he's about fifty feet above water. We call him the stack man. Nick's job is to untie one pot at a time, because if the boat lists suddenly, an unfastened pot will tumble overboard. He's exposed up there, with the wind howling and the stack rolling underfoot, and a lot more likely to fall overboard than if he were working on deck. So the stack man always wears a lifejacket or some sort of flotation device.

Meanwhile, Edgar is on the upper deck, behind the wheelhouse, operating the crane controls where he has a clear view of the lower deck. As Nick fastens the crane's hook onto the pot, Edgar hoists the pot into the air, moves it from the stern to the middle of the boat, and sets it down on the pot launcher, a steel rack that lays at a 45-degree angle against the rail of the boat. We call it the rack. Matt and Jake guide the massive pot onto the rack and unfasten the crane hook, which Edgar returns to the stack, where Nick should have another pot semi-untied and waiting.

Jake and Matt then unlatch the door of the pot. The door is one entire side of the cube, a seven-by-three panel that swings open. Once it's open, Jake squirms inside to bait it. This is the dirtiest job on the deck, and we usually reserve it for the greenhorn. The bait is a rank stew, usually chopped herring stuffed into nylon bags. We string up a whole codfish, too. Lying on his back, Jake ties the bait to the steel rails overhead. Once the bait is attached, he crawls out. Then, he and Matt close the door behind him, and Jake ties it shut with cord.

Meanwhile, Matt has reached into the pot and hauled out the shots and buoys stored inside. A shot is a coil of braided line 33 fathoms long. (Cowboys use rope; fishermen use line.) Since a fathom is six feet, the two shots combine to reach about four hundred feet. One end is tied to the pot, the other end is tied to the two buoys. The bottom shot is made of poly-fiber, which floats, keeping it off the bottom; the upper shot is nylon, which sinks, keeping the slack from floating to the surface where it can get tangled in the propeller. Matt sets the coils on deck so they won't get tangled with any of the machinery.

With the pot baited, and shots and buoys stacked on deck, it's time to launch. Norman is behind the wheelhouse on the lower deck, running the hydraulic controls of the pot launcher. At his command, the rack rises up and tilts over the rail, and the pot slides overboard and splashes into the water. Then Matt and Jake heave the shot and the buoys in after them.

This is perhaps the most dangerous moment of the whole process. When that pot drops overboard, it sinks straight to the bottom, trailing four hundred feet of line. If a deckhand gets tangled in that shot, or steps in a bight—that's what sailors call a loop of line—the weight of the sinking pot will immediately cinch around his ankle. In the next millisecond, the pot will pull that man overboard, and drag him toward the bottom of the sea. The men on deck have no way to save him. Even if we cut the line above him, the man is still underwater, tied to a seven-hundred-pound sinker. We can't dive deep enough to cut him loose. The man has only a minute or so to survive, and his only hope is to cut himself free, which is nearly impossible. Getting yanked off the deck will probably break an arm or leg or knock him unconscious.

For the sake of argument, let's say that even after getting tugged over the rail, falling two stories, and then plunging like a fishing lure toward the bottom of the Bering Sea, a fisherman still has his wits about him. He now has one chance. Every deckhand wears a knife on his belt. His one hope is to find that knife in the mess of line and raingear, pull it from its sheath, and saw through the taut bit of shot between him and the pot. If he can do that and swim back to the surface *and* if his boat can see him and throw him a life ring, then he might live. If it's dark, as it is eighteen hours a day in the northern winter, there's little chance that he'll be seen. Even in the best of circumstances, all of this has to happen in the four minutes that he can survive being in the water.

So that's how we set the pots. Bait. Launch. Repeat. We drop them in a "string," which doesn't actually mean that the pots are tied to one another. It just means that they are in a straight line, so that when we return to pull them, we won't have to change course to find them. It takes us about eighty seconds for each pot, so we can drop a string of thirty in about forty minutes. Depending on conditions and how far apart we set the strings, dropping the whole stack of two hundred pots will take between eight and fourteen hours.

It's important that we know how deep the water is where we set our gear. If we have 400 feet of line, and the water's 300 feet deep, there's no problem: the extra hundred feet of slack will just sink and drift underwater. But if the water's 500 feet deep, or even 405 feet deep, then the buoys will be pulled under by the weight of the pot, and we'll never see them again. That's an expensive mistake.

After we drop the pots we let them soak. Ideally, the crabs on the sea floor are so drawn to the bait that they climb right in. The size of

Nick and Norman guide the buoys through the block as
a pot is hoisted from the water. *(Courtesy of EVOL)*

the opening varies according to species: for king crab it's nine inches
by thirty-six inches. For opies it's smaller, so the king crab can't get in.
The crabs crawl through the opening, then fall to the bottom of the
pot, where they can't get out. In the web of each pot, a few strands of
cotton twine are sewn in with the nylon. This is an escape mecha-
nism. In the event that the pot is lost, the cotton will rot away, the
mesh will tear open, and the crabs will go free. Fish and Game began

Deckhands pull the pot of opilio crab onto the rack. *(Courtesy of EVOL)*

requiring this in the early nineties; most of the fishermen hated it because it required more work and rigging. It's the right thing to do, though. It has taught us to be more aware of ecology. The entire fleet has really gained more respect for the health of our fisheries.

The pots soak anywhere from a few hours to a few days. We pull a couple of sample pots to see if we're on the crab. If after a short soak the test pots show good signs of life, we're in business.

Then comes the real work: hauling the pots back onboard. I will have marked on my plotter where each pot is dropped, so we have a general idea. But knowing the GPS coordinates does not guarantee that when we return at night, in a snowstorm with 20-foot seas, I will actually see those two buoys bobbing on the water. So I maneuver as carefully as possible and scan the waves, sometimes with binoculars. Remember: the crab pot is not attached to the boat. The only way to get it onboard is to find and collect those buoys. So I'm just barely moving forward, letting the engine idle. If I miss the pot, I'll have to turn around and start all over.

Once I spot the buoys, I bring the starboard side of the boat as close as I can. It's not easy bringing a 125-foot boat alongside a buoy hardly bigger than a beach ball. If there's bad weather, I have to be careful to not let the boat get sideways to a huge wave. I slither through the waves like a snake. The throttle is important for this. Too much throttle, and the impact of the waves is greater. Too little throttle, and I lose control.

As we approach, Matt readies the grappling hook. It's a three-pronged steel hook, two feet long by ten inches across and weighing six pounds. As we idle forward, Matt chucks that hook between the two buoys, like threading a needle. If he misses by inches to the outside, or even if he hits the buoy with the hook and it bounces away, he has to pull the hook back on deck and try again, as the buoys recede behind us. Once he threads the needle, Matt pulls in the line, bringing the buoys with it.

With the buoys on deck, we need to pull the pot up. There are some pretty strong deckhands on crab boats, but none strong enough to pull pots by hand. For that task we've got a hydraulic block. Matt feeds the shot through the block, and Edgar hits the switch and starts reeling in the 400 feet of line. This creates another hazard: 70 fathoms of line spilling onto deck in big loops. If the block were to break, the pot would drop back to the ocean, and again, if someone got tangled in the line, they'd be gone. So we have to coil. In my dad's days, one of the deck jobs was to coil the shots by hand as soon as it passed through the block. Now we have a hydraulic coiler, which looks like a big steel oil drum. We feed the line in, and the machine coils it.

Once the pot reaches the block, it's dangling at the starboard, banging against the hull. Now it's time for Edgar to step up. The block can only pull the boat to the rail; it can't hoist it over the rail and on deck.

After the pot has been emptied, Matt *(right)* coils the shot and Norman stuffs the buoys back in the pot to store till next time. *(Courtesy of EVOL)*

For that task we have a stationary hydraulic picking boom with a steel hook hanging from the end of it. Now Jake reaches over the rail and hooks the frame of the pot with the picking hook. Using the boom, Edgar hoists the pot up over the rail. If the pot is filled with crab, it weighs almost a ton, and it's dangling overhead as the boat is bucking in the seas. It's like trying to run a backhoe during an earthquake.

To guide the pot onto the rack, Nick and Jake reach over the rail and grab hold. This is another dangerous moment: if the pot takes an unexpected swing, it can smash and easily kill them. If the guys reach too far and lose balance, or if the boat lists suddenly to starboard, they can fall overboard. Two years ago when we renovated the *Northwestern*, we raised the rail by one foot, to prevent this.

Eventually Nick and Jake get the pot onto the rack. Then to make sure it doesn't slip off, we lock it with a pair of steel clasps called the

Jake, Matt, and Edgar sort crab while Nick stacks the empties on the stern, as Norm operates the crane from the upper deck. *(Courtesy of EVOL)*

dogs. With the dogs locked, we don't have to worry about the pot sliding off the rack and crushing someone.

Next, we empty the pot. We slide the sorting table up against the mouth of the pot, swing the door open, and tip the pot at a steep angle on the rack. The crabs come tumbling out into a big mountain on the table.

If we want to keep fishing the same spot, we'll re-bait the pot and drop it again. If we're moving to other grounds, we'll pull the old bait and store the shots and buoys inside the pot. Then we hook the pot with the big crane, and hoist it to the stern where we stack all the empties and lash them together.

Next we have to sort the crab. With king crab, if we're getting a hundred per pot, we're having a once-in-a-lifetime run. An average of fifty is phenomenal. If we're averaging thirty, we can't complain.

Beautiful fishing. Since opies are so much smaller, we need to catch a lot more. Fifty opies is a miserable pot. We want to average two to four hundred, and an awesome string might have pots of seven hundred, even one thousand. Those numbers don't refer to what we haul on board, but to what we actually keep.

What we're looking for is large male crabs, but that's not always what we get. All females must be thrown back. We measure the width of the males' shells. King crab must be six inches wide, and opies must be four and a half. The little ones are thrown in the "shit chute" that sends them overboard. The legal ones are dumped through manholes into holding tanks. Each deckhand counts keepers, and when the table is empty, we combine our counts and report the total to the captain on the intercom.

Some deckhands compete to see who can sort crabs the fastest. Of course, you can't always be the fastest, but most deckhands will do their damnedest not to be the slowest. So hands and crabs are flying in a fury, measuring, sorting, tossing. It gets going very fast. If you're catching a lot of crab, it's an exciting time that raises spirits: You know you're getting paid for each crab you take home.

The *Northwestern* has three crab tanks: fore, middle, and aft. Together they hold 218,000 pounds of crab. Pumps keep the tanks filled with recirculating seawater, because live crab require fresh seawater until we get them to port. Before we catch crab, we might leave the tanks completely empty, unless we're in heavy seas, and then we might pump them completely full for extra ballast. What we don't want is a partially filled tank—a slack tank. As the boat lists in the seas, that water sloshes around, just like a cup of coffee on your dashboard when you make a quick turn. You have to remember that 30,000 gallons of water weights 120 tons, and the force of that water sloshing can capsize the boat. Think of a swimming pool, half empty. Now pick it up and shake it. That's the kind of force we're talking about. Water-filled tanks present another hazard to a deckhand: if he were to fall through the manhole into this cavern of 38-degree water, he could break his leg.

The process of setting gear, pulling pots, sorting crab, and stuffing the tank goes on around the clock, hundreds and hundreds of repetitions until the tanks are full. Two hundred thousand pounds of king crab in the holds is worth somewhere between $600,000 and a million, depending on what the canneries are paying and the market will bear. That same gross of opies pays around $300,000.

Once we get paid, we deduct from that gross the costs of running the trip: fuel, bait, food, landing tax, and crew airfare. Fuel is by far the largest expense. The longer the season, the more we use. In king crab season we typically burn about twenty-five thousand gallons of diesel. At three dollars a gallon that's seventy-five grand. Of course, we like to use as little fuel as possible, but sometimes economics dictates otherwise. If fuel costs a dollar less per gallon in Seattle than it does in Alaska, and we need fifty thousand gallons, it's well worth it to make a two-week round trip south to buy fuel. It saves us fifty grand.

So let's say we bank a million dollars of crab. The expenses might be $300,000. The remaining $700,000 is divided among crew and owners, with the crew taking 40 to 43 percent. Each deckhand gets 5 to 7 percent. The deckhands usually earn in the neighborhood of fifty grand for the one-month king crab season, and another fifty grand for the two-month opilio season. The rest of the profit goes back to the boat, of which Norm and Edgar and I are equal owners. Maintenance is expensive. In 2008, our overhaul cost more than $800,000. She's getting older, but now she's as good as new.

On the *Northwestern* we pay a higher crew share than most boats. That's how Dad always did it, and we follow his example. We want to get the best deckhands, and we want them to stick around. We want them to earn money. We want them to put that effort into the boat. We want them to be lifers. We treat them so they will come back, and when they do, that boat is automatically safer. Every boat has its own personality and way of doing things, and we teach our guys the way we have fine-tuned over the years. The more turnover you have on deck, the less people know the systems, the more accidents you'll have. That's almost guaranteed.

By the time I came to crab fishing, Dad and Karl were already established skippers. I had big shoes to fill, but at the same time their hard work opened doors for me that weren't open for others. So I want to tell the story of how they got their start in Alaska.

From 1961 to 1963, when Karl was fishing in Seattle, Sverre was stationed in Germany with the U.S. Army. He never talked about it much, and no one in the family knows much about that time of his life. We know that he spent some time in a tank—we have a picture of him in a Russian-style fur hat with ear flaps leaning up against a tank in a field of snow. We know that he learned to play guitar by hanging out with a Cajun friend named Charles Bosach who would later become a country-western singer. We also know that before Christmas in 1962, he wrote to his brother in Seattle from the army base in Germany. He asked Karl for two hundred dollars so that he could go home to Norway for Christmas. At the time, Karl was making good money—a thousand dollars a month—on a halibut trawler. He sent Sverre five hundred.

Instead of spending the holidays in the barracks in Germany, Sverre bought a ticket to Karmoy. When he arrived home in his U.S. Army uniform, he cut a striking figure, with dollars in his pocket and a few years of world travel behind him. He brought his mother a Blau-

Dad stationed in Germany, 1963. *(Courtesy of the Hansen Family)*

punkt turntable and a stack of Hank Snow records. The neighbors still remember how in warm weather she threw open the windows and let that twang sift out over the beach for all to hear: *I've been everywhere, man.*

When Sverre showed up at the town dance, he was provoked by a couple of guys from the next village over. He calmly slipped out of his uniform, stepped outside, and gave them a beating. Apparently he was handsome enough in that uniform to attract the attention of a fisherman's daughter, Snefryd Jakobsen. She was a pretty young girl from the same village as him. Sverre courted her and soon proposed. Uncle Karl likes to say that if he hadn't sent him that five hundred bucks, my father would never have met my mother—and we wouldn't even be here.

When his army tour ended, Sverre returned to Seattle. Now he had to earn enough money to bring his fiancée to America. He went down to Fishermen's Terminal and found Dan Luketa. He told his old boss about his girl back home, and his plan to bring her over. All he needed now was a spot on a good boat.

"We're all full," said Dan Luketa. Taking another crewman would mean a smaller slice of the profits for each. The crew was opposed to bringing Sverre back, but Luketa had made Sverre a promise. He kept his word.

"Boys, we're going one extra man," he told them. The rest of the crew grumbled, but Luketa didn't budge. Sverre would never forget this honorable decision by an honorable man.

The money was good, but not great. Sverre tried to buy a car but was denied a bank loan. With Uncle Jorgen gone, Karl and Sverre shared the basement in the house of a Norwegian couple, the Helgevaards, near Third Avenue in Ballard. Karl now had the old Ford his aunt had left behind. One night he and Sverre and the boys had been drinking at Malmen's and were headed up to another party. The car was parked out back.

"Let me drive," said Sverre.

"You can't drive," said Karl. "You don't even have a driver's license."

So Karl started up the car and the others piled in. He decided to show off a bit. He whipped around Ballard Avenue and sped up. The damn thing. He lost control and sideswiped a string of parked cars. The fifth car stopped him. A little Volkswagen was spun in a half circle and knocked up on the curb. The cops arrived. By then, they knew the Karmoy boys pretty well.

"Ah, just get them out of here," the cops said.

Karl paid a bit of money for the Volkswagen, and was let off without an arrest. One of the cops even sold Karl his big Chevy as a replacement.

The brothers were competitive. During Sverre's absence, Karl had established himself as one of the best young draggers in Seattle. They say he had the fastest hands for repairing nets of any man in the fleet. By age twenty-six, he was already a skipper, one of the youngest in the Northwest. On his first trip as captain of the *West Ness,* Karl took his crew up the Canadian coast to the Hecate Strait. He was a bit unsure of himself. There he recognized a boat call the *Tordensjold,* whose salty old captain, Carl Servold, was a friend. He raised him on the radio.

"How much fish are you getting?"

"It's good fishing here," said Servold.

"How many fathoms?" Karl said.

The old-timer answered in single syllables. Karl kept peppering him with questions. How strong is the tide? How long do you soak for? And on and on. Finally Servold had had enough.

"You're asking too many questions there, young feller."

Karl hung up the radio and got to work, dropping his nets out near the rest of the fleet in deep water. While the crew was changing the net, Karl climbed down from the wheelhouse to help them. He forgot to turn off the fathometer. When he returned to the bridge, the boat had drifted far toward the bank, in just 30 fathoms of water—or about 180 feet. The fathometer had spewed its paper readings all over the floor. Karl took a look. Big bumps of fish everywhere. *Man, there are a lot of fish here!* So he dragged his net back and forth over the shallows for a few hours, and came up with forty thousand pounds of snapper, perch, cod, and sole. He did it again. Same thing. In the morning he returned to his honey hole and did it again. His ship carried 200,000 pounds, and at this rate he'd fill up in three days. Finally, the other captains in the fleet got curious. Carl Servold crackled through on the radio.

"What the hell you doing down there?"

"I'm fishing," Karl said.

"You getting any fish?"

"Yeah, I'm getting fish."

On the third morning the fishing was just as good. Now Servold called again, with all sorts of questions. How much fish are you pulling? How deep are your nets? A cocky kid with 200,000 pounds in the tank, Karl couldn't resist.

"You're asking too many questions there, old-timer."

Karl pulled the nets and steamed toward Seattle, jogging a bit off course to pass close to the other ships. They assumed he'd given up on his shallows and had come to fish with them. Of course, he didn't stop, and just kept motoring south. Servold came through on the radio, calling to Karl's stern.

"Where the hell you going now?" he demanded. The *Tordensjold* held only a third as much as the *West Ness*, and still it wasn't full.

"Don't you go home when you're loaded?" Karl said.

"*Jesus Christ!*" said the old-timer. The whole fleet had been out-fished by a rookie.

Even with Karl's success, Sverre didn't ask him for a job. He was too proud to beg a favor from his younger brother. He wanted to find his own way. In the spring of 1963 Sverre got a job on the wooden schooner *Seattle* longlining for halibut. As the greenhorn, he was also cook. One of his crewmates remembers Sverre as "desperately in love" with his girlfriend back in Karmoy. "He was very quiet and very hardworking," says Knut Thorkildsen. "He never talked bad about others. We knew he was very much in love with her. He was thinking about her all the time he was fishing."

In April they steamed north to Alaska for the opener off Kodiak Island. Sverre saw the big north for the first time—the steaming volcanoes, brilliant glaciers, and jagged peaks cresting up out of the gulf. During a storm the *Seattle* took shelter in Akutan. As they anchored in the bay to re-bait their gear, a teenage kid just a few years younger than Sverre rowed up in a skiff. His name was Charlie McGlashan, an Aleut whose family had settled the island a century before. Sverre introduced himself. "We call him the Cabin Boy," snorted the skipper, sure to embarrass Sverre whenever he could.

The young men became friends. Charlie remembers, "Sverre was gung ho, working and trying to prove himself among the older guys."

Charlie's family owned the general store, and a rough bar called the Roadhouse. He traced his Alaskan roots to his great-grandfather, who had stowed away aboard the SS *Patterdale* in Liverpool, England, in the 1870s, jumped ship in San Francisco, then made his way to Alaska to hunt seals. In 1878, he founded the village of Akutan, married a woman from the island of Attu, and had thirteen children. His descendents have been there ever since. There were a few native Aleuts elsewhere on the island who lived in barabaras, traditional submerged mud dwellings having sod roofs. Charlie's grandfather had the first codfish station in the Aleutian Islands. In 1912, Norwegians arrived and started a whaling station that lasted until World War II. When the war broke out, all villagers on Akutan and the rest of the Aleutian Islands were shipped to U.S. internment camps near Ketchikan, where 10 percent of them died—and where Charlie was born.

Sverre and Charlie McGlashan would become lifelong friends. Over the years Charlie came to visit Sverre in Seattle, and Sverre would lend him money when he needed it. Sverre would come to regard Akutan as his adopted home. He preferred it to Dutch Harbor. He would trade crab with the natives for their home-smoked salmon or their salmonberry jam. Their friendship even outlasted Sverre's life: Now with Dad gone, Charlie is like an uncle to Norman and Edgar and me. My path as a skipper would have been much different if not for their chance meeting.

Finally in January of 1964 Sverre's new bride arrived. Sverre took her for a Sunday drive in Karl's Chevy, all the way up to Snoqualmie Pass, where she stumbled around the snow in her high heels. Four days later they were married at the Rock of Ages Lutheran Church in Ballard. A reception followed at the Helgevaards' home. All the Karmoy fishermen they knew were invited. Sverre asked his friend Soren Sorenson to take pictures, but Soren said he didn't know how to use a camera. So Sverre loaded the film and gave basic instructions. Soren snapped photos throughout the ceremony, only to discover later that the film hadn't been properly attached to the spool. The roll was blank. Karl missed the wedding. He was in Alaska.

The newlyweds rented an apartment in a brick building at Third Avenue and 63rd Street in Ballard, a modern-looking place with a nice kitchen and big windows. Tormod Kristensen and his wife lived in the same building. Karl kept a bedroom, but he was hardly there. He stashed some clothes while he was at sea.

Sverre and Luketa had been dragging mostly around Puget Sound and just up the coast in British Columbia. Now that fishery was depleted. Dan Luketa was getting old and blind, and didn't want to learn the new skill of crab fishing, but he wanted to send his boats north, to see if they could get a chunk of the loot that crabbers had been winning. Luketa converted the *Western Flyer* for crab pots, and hired a young Alaskan named Howard Carlough to run it. Carlough was a pioneer, a native of Seldovia who began crab fishing in 1953 when he got out of the army. "He'd heard about me killing those crab," says Carlough. "He asked me if I'd come down, and he made me some deals I couldn't refuse." For crew, they hired two experienced crabbers named Bill Osborne and Jim Markey. Osborne was a Ballard native who ran off to join the navy at age seventeen, fought in eight naval battles in the Pacific during World War II, and witnessed the first postwar atomic bomb test at Bikini Atoll. They still needed a greenhorn—and a cook.

"I got a young guy who would do it," said Dan Luketa. "A hell of a kid."

Sverre Hansen landed his first job on a crab boat. He was headed north to the Bering Sea.

5

LET THE KID RUN THE BOAT

Two hours into fighting the fire, Sverre considered—for the first time—abandoning ship. Maybe bringing the *Foremost* to the Bering Sea had been a mistake. It wasn't designed for these conditions. Now he was paying the price for taking his boat past its limitations.

Abandoning ship was a terrible decision to have to make. They didn't have the thick rubber survival suits that are now standard issue. They had life rings, but in the bitter December waters, Sverre knew a man had only four minutes. Their only hope was the eight-foot rubber raft and the skiffs lashed to the wheelhouse. With no radio contact, there was no guarantee they'd be sighted by another ship. No one knew where the *Foremost* was, and if she went down, no one would come looking. With the storm blowing in, most of the fleet was laying in port in Dutch Harbor. The raft might simply drift out to sea.

Leif Hagen, who had discovered the life raft buried in the galley just a few days before, now took ownership of the thing. Retreating from the firefight, he pulled the raft from the wooden boxes behind the cabin and pulled the cord. The raft sprang to life, a giant red donut covered by a red rubber tent. The men heaved it over the rail and it splatted in the turbulent water, tethered to the boat with a single strand of line. Hardly the thing you'd want to trust your life to.

It was now or never. Sverre ordered the men to ready the skiffs. They began lowering the wooden boats and cutting the cords. Icicles still clung to the bottoms of the skiffs. These looked more seaworthy than the raft, anyway.

An explosion ripped beneath the boat and the deck shuddered. The men instinctively ducked and covered their heads with their hands. Maybe each man saw a vision flicker through his head: a childhood memory, his mother's embrace, a kiss from the girl he loved. In the next instant they might be engulfed in fire. Slowly, they raised their heads. They were still alive. The boat was still intact and afloat.

Suddenly there was another explosion. The wooden deck thumped beneath them. Then came another blast, like sticks of dynamite exploding below.

"It's the batteries!" yelled Leif.

He was right. Their fluids had been boiling and one by one they burst open like grenades of sulfuric acid and lead and hard plastic.

Boom! Boom! Boom! Another three exploded. How long did they have before the real bombs—the oxygen tanks—ignited? How long before the fuel tank erupted?

"Let's get out of here!" cried Krist.

A blast of heat hit Sverre's face. The men gave up on the skiffs and bolted for the life raft. Sverre looked at the raft as Leif and Magne raced to untie it. He gave the order, "Abandon ship!"

———

In 1988, Chris Aris was just out of high school, living with his parents in a suburb north of Seattle, not sure what to do with himself. Hanging out at a local tavern, he met a guy from the neighborhood, Mark Peterson, who was just a year older. Sometimes after the bar closed, the party would move a few blocks to Mark's house. The first thing Chris noticed was that Mark drove a Porsche and owned his own house. Mark told him that he had a job crabfishing in Alaska on a boat called the *Northwestern*. Chris Aris was blown away. He had no idea what crab fishing entailed, and had never been to Alaska. When his friend asked if he wanted some work on the docks that summer, he jumped at it.

Aris worked for us down in Ballard, rigging pots and painting the boat, whatever odds and ends needed to be done. "You gotta tie that pot down faster if you want to go out with us," someone would tell him. Aris was already tying as fast as he could, but he tried harder. I liked that. He came onboard with zero experience but proved himself quickly. "It was a bit more difficult for me to learn because I'm left-handed," he says, "and everything on the boat was set up for right-handers. So they'd teach me to tie a knot and I'd have to figure out my own way." Aris was a hard worker and we all liked him. He had blond hair, blue eyes, and a square head, but as it turned out he wasn't Norwegian. Even so, the next summer we hired him to work on deck.

Our plan was to get to Alaska early, before the fall king crab season, and go out west to Adak Island for brown crab. In July we all got onboard in Ballard, passed through the locks, and motored north. Mark Peterson took that trip off, so the rest of the crew was Pete Evanson, Brad Parker, Steiner Mannes, and Edgar, who was eighteen and just starting working on deck. The oldest men on board were Evanson and Parker, who we called "the old man." They weren't even thirty. I was twenty-three.

Aris the greenhorn was sick for the whole seven days motoring across the gulf. He just lay in his bunk puking. At one point he thought some Rolaids might help his stomach, but that just made him puke foam. He felt horrible. I'd been there before. You just stare at other people eating and wonder how they can possibly do it. You're hungry, but the last thing you can do is eat. It's the hangover that never ends.

I had been relief captain on the *Northwestern* for a few trips here and there, but this was my first season as full-time captain. I'd been given my first shot the previous year, when Tormod had taken a break. Tormod had been grumbling a bit about teaching me the ropes of being skipper, and I don't blame him. "Why should I teach him all my tricks when he's just going to take my job?" he muttered to a deckhand. Meanwhile, Mangor Ferkingstad, my old mentor, was the deck boss, and he was looking to become skipper. He was blunt about it. He went to my dad.

"Either you let me take it," Mangor said, "or you let the kid run the boat."

It was the furthest thing from the Old Man's mind, but once Mangor mentioned it, I guess my dad liked the idea. At the time I was home on a break, sick with the flu. Dad came into my room and was mumbling about something.

"What's wrong?" I said.

"You want to take the boat?"

I didn't really realize what he was asking me.

"Yeah, yeah, whatever."

"You're going to take the boat."

So then I flew up there, thinking he was going to take me out and show me what to do. The last thing I expected was to become captain. It was opilio season. My dad had always thought of opilios as trash, a tiny inedible bug that had to be thrown back when they happened into

Dad *(left)* aboard the *Western Flyer*, his first time crab fishing in Alaska, 1965. *(Courtesy of the Hansen Family)*

Norman *(right)* and I behind grandmother's house in Karmoy, 1971. *(Courtesy of the Hansen Family)*

Hansen Family Christmas, 1972: Uncle Karl, Dad, Norman, I, Edgar, and my cousins Jan Eivin, and Stan. *(Courtesy of the Hansen Family)*

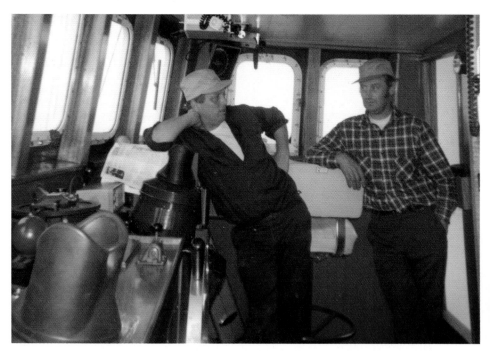

Dad *(left)* and Tormod Kristensen aboard the steel *Foremost*, 1973. *(Courtesy of the Hansen Family)*

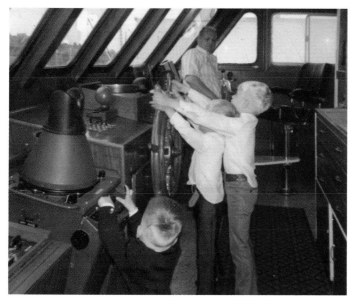

Dad with Edgar, Norman, and I aboard the steel *Foremost*, 1973. *(Courtesy of the Hansen Family)*

Dad oversees the crew pulling on a pot on the steel *Foremost*, 1973. *(Courtesy of the Hansen Family)*

Here I *(left)* am with Edgar *(rear)* and Norman fishing on the lake. *(Courtesy of the Hansen Family)*

Family portrait, midseventies. *(Courtesy of the Hansen Family)*

Here I am as deckhand, Edgar as captain, circa 1980. *(Courtesy of the Hansen Family)*

Norman as a deckhand on a Bristol Bay salmon boat.
(Courtesy of the Hansen Family)

Here I'm shown with a pair of good salmon on the *Jennifer B*, 1980. *(Courtesy of the Hansen Family)*

I'm working as a deckhand on the *Northwestern*, fishing for crab, age fifteen. *(Courtesy of the Hansen Family)*

I am shown with an Aleut family in Nome, circa 1982. *(Courtesy of the Hansen Family)*

Mom and Dad fishing on the lake. *(Courtesy of the Hansen Family)*

Thirty-three years of family pride—and still going strong. *(Courtesy of EVOL)*

I'm on deck with Jake Anderson, Matt Bradley, and Edgar. *(Courtesy of EVOL)*

Norman Hansen. *(Courtesy of EVOL)*

Edgar Hansen. *(Courtesy of EVOL)*

your traps because they took up the valuable space needed for king crab. He ran for miles to get out of them. To fish them on purpose was degrading work. As an old-school king crabber, he thought opilios were beneath him.

Dad flew up to Dutch after me and met up with Magne Nes. By this time, these guys were all considered old-timers. I was getting ready for the trip, not quite sure what to do. And then Magne told Dad, "Just leave the guy alone, Sverre! Let him do his own thing."

Then it hit me. My Old Man was seriously expecting me to go out as captain without him.

It was May. The weather was nice. One of dad's friends, Walt Christensen, jumped on the boat docked next to me. He gave me a blank bearing book—to record my fishing positions—and said this is how I do it. I followed Walt out to the grounds. When we got close to the rest of the fleet, he radioed and said, "You're on your own."

I had been watching Tormod and my dad, so I had a good idea of how to navigate and set gear. Now I just had to figure out where to set it. I had a radio. Channel 12 was the squarehead channel. The Norwegians always had their own channel. When we were doing watch at three in the morning, we could always find some other squarehead to talk to. "Anybody on there?" He'd be on another boat, some Norwegian guy, a crew member. We were always communicating.

The fleet understood it was my first time, so they hazed me. By then, Mangor had been hired as skipper of the *Western Viking*, where he did very well. I had to radio him and ask him how to punch in waypoints on the loran—Long Range Aid to Navigation—which is what we use to find our pots once we've set them. The others knew I was having a hard time finding the crab. There was a learning curve for me, and the crew had to suffer through it. My strings of pots were a mess. They were incorrectly spaced, jumbled on one another. I got tangled in my own gear or ran over the buoys with the propeller. I also seemed to come down with dyslexia. I'd miswrite the number of a pot, so after wasting time searching for #89 I'd find #98 instead.

When I did get on a really decent pile of crab, the crew could right away hear it in my voice on the radio. They knew I was on the crab, and they were all laughing. "Sounds like you found a pile!" Oddvar said on the radio. I was so proud to hear this praise from him, someone I'd looked up to for a decade. I was happy. I was having a blast. But I still didn't feel like a real skipper—I felt more like a dumb kid sitting in a chair pulling pots. At the time, I was twenty-two and may have been the youngest skipper in the fleet.

There *were* a few younger captains, but in my mind most were old-timers. If a boat is under two hundred gross tons, the captain doesn't need a license. The only thing determining whether I could be captain of a vessel the size of the *Northwestern* was if the insurance company would agree to insure me. It was very subjective. If the pool managers thought you were good enough, they would let you in. Otherwise they'd blackball you. Many years later I learned that some of the pool managers didn't think I had what it took. In fact, one guy in the company had made a bet that I was going to mess up and hurt someone. Years later he revealed this secret to me—and admitted that he'd actually lost money betting on me, because I'd survived those early years without an accident.

Eventually I created my own way of doing things. I got to know my boat and crew to where they were like an extension of myself. I was obsessed with numbers. The crew called me Captain Casio, because I was always plugging numbers into my calculator. If we could pull X amount of crab in Y minutes, then we were earning Z dollars per hour. One time, a skipper in the fleet announced that he was taking the day off to go to port to watch the Super Bowl. I was stunned. My crew was looking at me, wondering if we should do the same. I whipped out my calculator. So, four hours to watch a game?

"Well, four hours of crab, that's ten grand," I said. "You want to miss that kind of money? That's the way it works."

Pulling a full pot is the best feeling in the world. But being a psycho, I always get depressed because I *know* the next time it won't be as full. That thought bothers me. I should be happy, but I always want more. I never looked at fishing primarily as fun. It's work. I tried sportfishing for marlin in Mexico once when I was a kid, but I didn't like it. Some damn dentist caught the thing instead of me.

Anyway, there we were in the summer of '88, motoring north, for my first full season as captain. We were all relieved when we reached Akutan. We called Charlie McGlashan on the radio and he came down to the docks to welcome us. Charlie Chan from Akutan was his nickname. He was one of my father's best friends, like family. We cooked up a platter of *kumla*, Norwegian-style potato balls—his favorite. It's a poor man's tradition that goes back generations—we mix the potatoes with rye and boil them for an hour, then add salted lamb, carrots, and rutabaga, and pour melted butter over it. It's hearty and bulky. The old people say if you eat it, don't go swimming—you'll sink. We learned to cook it from our parents when we were kids. We took the food up to Charlie's house. It was always good to see a familiar face. After dinner, he took us up to the Roadhouse for a beer. His aunt owned the place.

Akutan hasn't changed much since my dad's first trips there in the early sixties. Trident Seafoods replaced the old floating processing ship with a new cannery, which employs up to eight hundred workers in peak season. Those workers live on the west end of the harbor. The village on the east end is relatively untouched by modernity, still only about seventy-five people living there year-round. Over the years, the

The Russian Orthodox church in Akutan. *(Courtesy of EVOL)*

Aleuts intermarried with the Scots and Russians. The natives moved out of the barabaras and into the village, and eventually formed a corporation that owns much of the land and businesses.

Akutan makes Dutch Harbor look like a big sophisticated city. Because the terrain is so steep, there is no airstrip, and the only way in beside boat is the Grumman Goose, an eight-seat amphibious airplane that flies from Dutch Harbor when the weather is good. My brothers and I prefer fishing out of Akutan because there's really nothing to do but fish, and we can stay focused. I like working as much as I can, with no one telling me when to stop. Dutch Harbor has the bars and people and all sorts of distractions. Since Akutan has traditionally been my family's hub of operations, we don't go to Dutch as often, so it's easy to avoid the bars. When we finally do get out in Dutch, we tend to go overboard.

The night we had dinner with Charlie in Akutan, we had just missed the big excitement. The night before, two cannery workers had fought in the bar. One pulled a knife and stabbed the other. Killed him. As Charlie told us the story, Aris the greenhorn looked at me. I know he was thinking, *What am I getting myself into?* Looking back now, the murder was probably an omen of bad luck to come.

From Akutan we steamed for Adak, about four hundred miles west. It's way out there. Some seasons have been closed in Adak, and not too many fishermen of the new generation have been out there.

We had Aris filling bait jugs. As a greenhorn he didn't get a crew share—we paid him a flat hundred dollars a day. Of course we hazed the greenhorn, because that's how it goes. "Hurry up! Get over there! Come over here!" One time I went down to the galley while the crew was taking a five-minute break. Aris was slumped at the table, face in his hands.

"What do you think of crab fishing, Aris?" I said.

"It's a living hell."

I laughed. "Okay, five minutes is up."

Almost as soon as it started, the Adak trip was a disaster. Instead of the single pot fishing that we normally do, we were experimenting with longlining, in which all the pots were attached to a single line and hoisted in by a much larger power block. Out west near Adak, in the eastern part of the Bering Sea where the seafloor is flat, you're fishing above underwater volcanic cliffs. As we were dropping the line of pots, the fathometer was reading 120 fathoms. The problem was, I was getting a double-echo off the bottom: the water was actually 400 fathoms deep—2,400 feet. In other words, I had dropped two strings of twenty pots nearly a half-mile underwater, and had no way to retrieve them. I didn't know what the hell to do. It was as if I'd thrown them off a cliff. My stomach clenched. I felt like I'd been kicked. This was more than fifty thousand dollars' worth of pots stranded on the bottom of the ocean. I was screwed. I went down and told the crew what I'd done. They glared at me. No one knew how to get the pots back. They weren't pleased to be working for a rookie captain.

You have to remember that I'd been working all my life to get where I was at that moment. I was finally skipper of the *Northwestern*. I had stepped out from Dad's shadow, proving I had what it takes to run a crab boat. The next thing you know, I'd made such a mess of things that I had to swallow my pride and call for advice from the very last person I wanted to talk to: the Old Man. On a clear night our single sideband radio would connect directly to his house in Seattle.

"You did *vhat?*" My dad understood English perfectly, even on a radiophone. He had this habit, though: if you told him that you screwed up, he'd make you repeat it, as if he didn't understand. He just liked to see a guy squirm.

"I dropped the pots at four hundred fathoms."

"Four hundred fathoms?"

"Yeah."

"How many fathoms?"

"Four. Zero. Zero."

"You can't fish four hundred fathoms, you dummy!" he said jokingly. As he had gotten older, Dad had adopted *dummy* as one of his favorite words. I think he picked it up from watching *Sanford and Son*.

"I know that, Dad."

"There's no crab at four hundred fathoms. And you'll never get the pots back."

"Yeah, Dad."

"You can fish *fifty* fathoms. You can maybe fish *one hundred* fathoms, but you can *never* fish *four hundred* fathoms!"

I stared at the radiophone. *No fucking shit, Dad.*

We ended up dropping a drag line from the block. At the end of the line we attached a big metal hook. We knew approximately where the string was, and dragged the hook back and forth over the ocean floor, hoping to snag the longline. We did this for hours. I was pissed. Everyone on deck was pissed. We could just feel the money pouring out of our pockets, not catching a damn crab.

Finally the guys let out a yell. The nylon line tightened like a cable. We'd snagged it. We started to reel in the line. Unfortunately we'd hooked it more or less in the middle, so once it surfaced, it was a jumbled mess with pots hanging off either side, as if you'd picked up a railroad train in the middle. After nine hours of hauling and hassling, we salvaged most of the twenty pots, but not all. We had to cut a few loose and let them sink. So already we were in the hole. We owed money to the boat—that is, to my Old Man—which would be subtracted from our crew shares.

We started fishing. We were dropping gear two islands to the west of Adak, off Tanaga Island. The seas were calm, but the fog was thick and there were just a few boats fishing. It was hard to see what we were doing. When the string was dropped, hundreds of feet of line were floating beside the boat. I called out on the intercom, "All right, am I clear to go ahead?" They give me the signal, and I hit the throttle, but when the propeller started spinning it sucked the line into its blades in a big tangle. The main engine died. I knew right away that we were screwed: The nylon line was wrapped around the prop, heated up, and melted into a solid ball.

So there we were in the middle of nowhere, with no engine and no way to get the prop untangled. There were just a handful of other boats spread over the many miles of islands. I got on the radio and pretty soon another vessel came over and towed us into Tanaga Bay. There was nobody to help us—no towns, no mechanics, no people—but at least we could drop anchor in a protected bay until we figured out what to do next. We sat there for a couple of days. The greenhorn Chris Aris celebrated his twenty-first birthday moored in Tanaga Bay. Finally I was forced to call Dad—again.

"You did *vhat?*"

"Got a line wrapped in the propeller."

"You dummy!" he laughed. "That will kill the main engine."

"I know that, Dad."

"Why did you run over the line?"

Yeah, yeah, yeah. I hung up and rolled my eyes.

Finally the crew and I came up with a plan—if you could call it that. A nearby boat delivered a scuba tank and a wetsuit. None of us knew how to dive, but Brad Parker volunteered to go first. The water was freezing, and the old wetsuit didn't look thick enough to keep him warm. So we had a brilliant idea—one that we'd seen in *Escape From Alcatraz*. Parker stripped down to his shorts, we smeared him with butter, then wrapped him in plastic wrap straight from the kitchen drawer. We were laughing hysterically. It was like wrapping up an appetizer to put in the fridge. Then he put on a layer of long underwear over the plastic wrap and the wetsuit over that. He heaved the tank on his back and strapped a knife to his waist. The tank of oxygen was supposed to last forty-five minutes, which seemed long enough to get the job done. We hoisted him on the crane, swung him out, and lowered him into the water.

None of us knew how a wetsuit worked: It uses your body heat to warm the thin layer of water between your skin and the rubber. As it turned out, the plastic wrap trapped the body heat in Parker's body, so it couldn't warm the water. When he let go of the line and took a couple of strokes toward the stern he freaked out. The ice-cold water shocked his system, and he got claustrophobic in the suit of butter and plastic wrap. He dog-paddled back to the crane and we hauled him up. We stripped him out of the wet gear and wrapped him in dry clothes.

"That sucked," he said. We started thinking up Plan B.

An hour later someone took a look at the scuba tank. In the chaos, the valve must have been left open. All the oxygen was gone. Another day passed. No oxygen. No diving. No motor. No crab. No money. Then we had an idea—the diver could breathe from the air compressor hose in the engine room. I called the guys who had lent us the tank and asked what they thought.

"Ah, that's not really good air," they said. "And besides, it not pure oxygen, it's just air."

We didn't see any other option. I put on the wetsuit—without butter or plastic wrap. We unrolled a hundred feet of rubber tube from the air compressor, inserted it into the scuba mask, and duct taped the whole assembly to my head. They lowered me off the crane into the skiff, and then I plopped in the water. Fuck it was cold. But I didn't have time to think about it. I dove underwater and swam toward the prop. The line was just a mess, a hundred fathoms or more of it, snagged up in there like a bird's nest. I started hacking away with the knife. The air in my mask tasted like oil. It was nasty air from the machinery. I was hanging onto the propeller underneath the boat, and it felt like the boat was jumping up and down twenty feet. But the water was actually flat calm. Then I realized I was getting high from the fumes I was breathing! I collected myself, and then I started hacking away at the melted ball of line with the knife. It took a long time but I got it free. I rushed up to the surface. I got on the crane and they hoisted me up. I was shivering and cold and my hands were blue, but I was happy to be on deck.

"Give me a smoke," I said. We started the engine and motored back to sea.

On that trip we also learned that just because we could *fish* didn't mean we could *catch*. We were pulling blanks. It was horrible. Long-lining crab pots is a cluster fuck. These days, fishermen have perfected efficient ways of doing it, but back then we were experimenting. As we pulled these twenty pots up, we had to coil the line that connects them, and we ended up with stacks of line everywhere on deck. When it was time to drop the string, some of those coils were just sitting there un-attached, while others were "live"—that is, attached to the string and extremely dangerous once those pots went overboard. It was confusing and frustrating, especially when the pots came up empty.

We were hauling around the clock. During a short break for break-fast we were just exhausted. We had a smoke and coffee and just sat around the table for a quick break. I leaned my head on the table for a little rest, and passed out right there. The rest of the crew also fell asleep and then, waking, crawled under the table and crept off to their bunks for a nap. I woke up six hours later, alone at the table, with my face in a plate of scrambled eggs.

That night we were working under the sodium lights, pushing it as fast as we could to make up for all the time we'd lost. Steiner Mannes was one of the deckhands. He was only eighteen at the time, but defi-nitely not a greenhorn. Steiner came from a fishing family—his dad was Borge, another of my dad's good friends, and his brother was

Johan, who I'd fished with for years. Steiner had been fishing in the summers when he was in school. He was a good fisherman and made a full manshare. He knew what he was doing.

As we dropped the pots, Steiner nudged one of the coils with his boot to get it out of the way, thinking it wasn't live; but it was. The line tightened around his boot. There was a moment—a split second—where he looked down and realized what he'd done, but before he could react, the line cinched and knocked him off his feet. The string of pots was sinking behind the boat, and it yarded him toward the stern, like he was tied to a bumper and getting dragged behind a car. He was sliding on his back, yanked by the leg; the only thing between him and death was the three-foot rail that kept him onboard. The deck was clear of pots, which meant he had about one hundred feet to get dragged before he went overboard—and just a few seconds to figure something out.

Brad Parker chased him down the deck, hacking at the shot with a knife. He was sawing at the point where the cord was bound over the rail, but he couldn't cut all the way through. As Steiner approached the rail on the stern, there were only a few seconds before the inevitable. He would slam against the rail, maybe break his legs as he pushed against the line, and then get hauled overboard.

He came bouncing toward the rail, and with his last bit of strength braced his boots against the steel wall. The shot tightened and cinched down on his boot, but by some miracle, his foot wiggled free. The shot pulled his rubber boot to sea, but Steiner was left on deck, barefoot and breathless, stunned that he was still alive. Parker's knife was rubbed to a nub.

Steiner was incredibly lucky. It was the closest we've ever come to losing a man on the *Northwestern*. I learned that the mistake I had made was trying to set the gear too quickly. What could have been a tragedy was, instead, just the worst crab-fishing trip we'd ever had. "I remember I didn't want to remember it," says Edgar, "because it was probably one of the most god-awful seasons I've ever been through in my life."

By the time we unloaded our meager catch of brown crab, we had lost so much gear, soaked so much bait, and burned so much diesel, that we owed money to the boat, instead of the other way around. When the woman from Fish and Game inspected our logbook while we were unloading, she couldn't believe that we'd pulled thousands of pots—and had so little to show for it. She simply didn't believe me. Of all the men on board that trip, the person who earned the most money was the one guy who'd been guaranteed a wage. The greenhorn Chris Aris walked away with a hundred dollars per day, while the rest of us *dummies* got zilch.

As I learned in Adak, knowing how to run the boat and having a top-notch crew dropping pots doesn't mean you'll catch crab. It's like Tormod used to say, "It's not how many pots you pull, it's how many you have in the pot." So assuming you have a good deck boss like Edgar and a good crew, the captain's job isn't to fish. The captain's job is to find the crab. I want to explain how I do that.

Most commercial fishermen use computerized fishfinders that detect the bubble of air that fish hold in their gills. Their monitor shows them the cod, sole, or other fish below the surface so you know where to drop their nets. Crab fishing is fundamentally different: Crab don't hold any air, so they can't be detected. We have no way of knowing whether any crab are crawling around four hundred feet beneath us. Despite all of our technology and experience, ultimately we're merely dropping a steel box to the ocean floor and hoping the crab crawl in.

So while heaving pots overboard may not require as much skill as seine netting or trawling, in my mind finding the crab is more of a challenge—a gamble—which is why I like it.

"Sooner or later the blind sow pulled an acorn" is how the high-liner captain Bart Eaton puts it. "You didn't have to be a genius. If you pulled a pot and it was full of crab, you were in the right spot. If it wasn't, you must be wrong—you gotta move."

He makes a good point—if you're a rookie, or a short-timer—but if you plan to do this work for thirty or fifty years, you need a better system. So finding crab is a combination of experience, science, cunning, instinct, and luck. No matter how long you do it, there's always guesswork, and trial and error. The day someone invents a device that detects crab on the ocean floor is the day traditional crab fishermen like us may be out of a job. All of our years of learning the crab grounds and predicting their movement would be replaced by a computer.

Before I get to the nuts and bolts of crab hunting, let's figure out who, exactly, we're hunting. Crabs are prehistoric creatures that somehow found their way into the modern world. Their fossils date back to the Jurassic era, and the horseshoe crab has been around for 400 million years. It's a living fossil.

Humans have been eating crabs for as long as we could catch them. The animals haunt our oldest stories. In Greek mythology, the hero Hercules was battling Hydra, the many-headed monstress. The goddess Hera was rooting for Hydra, so she dispatched Cancer the crab to interfere. The crab locked onto Hercules' toe with his claw, but the hero wasn't fazed. He crushed the bug underfoot. Hera felt bad that Cancer got a raw deal, so she assigned him a permanent place in the heavens, which is how we got the zodiac sign Cancer, the crab.

Though we talk about "fishing" for crabs, a crab is actually not a fish. It's a bug, a closer cousin to a spider, scorpion, beetle, and centipede than it is to a salmon, or even a clam. Crabs are arthropods, the phylum of invertebrate animals with an exoskeleton such as insects and spiders. Fish, on the other hand, belong to the phylum Chordata, vertebrates that include birds, reptiles, amphibians, as well as human beings and all mammals. This broad phylum contains not only trout, salmon, shark, and halibut, but also sea mammals like dolphins, whales, and seals. Many other sea creatures are from the phylum Mollusca, or mollusks, which include clams, oysters, scallops, mussels, octopus, and squid. These animals are even more primitive than arthropods. Many don't have a brain, just a simple nervous system—sort of like the men who make a living fishing them.

Among the Arthropoda, a name that comes from the Greek for "jointed foot," the crab's closest relatives are its fellow crustaceans: lobster, shrimp, prawns, and crayfish. These beasts are of the class Malacostraca, meaning "soft shell," and of the order Decapoda, or "ten-footed." Crabs then break into their own infraorder of Brachyura, meaning "short tail." From there, crabs are divided into 93 families and 6,793 species, some of which live in salt water, others in fresh water. All crabs walk on eight legs, and have a set of symmetrical claws for catching, crushing, and eating their prey. They range in size from the few millimeters of the pea crab, to the Japanese spider crab, which has a leg span of up to thirteen feet, and can weigh more than forty pounds.

The best-known species in America is blue crab, the soft-shell variety famous on the Chesapeake Bay that you'll see live in tanks in grocery stores and seafood restaurants. They have been harvested all along the eastern seaboard for centuries. A smaller species, Dungeness crab, about eight inches across, is caught along the Pacific Coast. The larger snow crab, of the genus *Chionoecetes,* or "snow inhabitant" is native to Alaska and Russia, and is a delicacy in Japan and the United States. In other parts of the world they are called queen crab or spider crab. The two main types of snow crab are tanner and opilio. Usually if you go to buy either in a store, it will be labeled snow crab or Alaskan snow crab.

The crab that I most like to catch are scientifically not crab at all. King crab is not in the "short tail" infraorder, but rather in the "different tail" infraorder or Anomura. The main difference between king crabs and other crabs is that while classified as a decapod, the king crab actually has eight functional legs. It uses three pairs for walking and has one set of claws. Its fifth pair of legs is tiny and hidden under its body, or carapace. The king crab is much bigger than most "true crabs." It can measure as much as eight feet across and weigh anywhere from seven to twenty pounds. Its legs can be as thick as a child's wrist, and its claws are different than other crabs: A smaller feeder claw is used for manipulating and carrying food, while the big killer claw crunches its victims.

There are about forty species of king crab, but for commercial fishing, there are basically three varieties: red, blue, and brown. The best-known is the red king crab, *Paralithodes camtschaticus.* The Latin name means "for fighting," a reference to its unusually large claws, and "Kamchatka," the part of Russia bordering the Bering Sea. The reds have red- or white-tipped spines, and their meat is pinkish. We fish them out of Dutch Harbor in Bristol Bay and in the Bering Sea. The season occurs in the fall—it's the one that you see on the first half of the *Deadliest Catch* season.

We catch blue king crab, *Paralithodes platypus,* in the northern

Norman at age six, and Edgar at age thirty-seven, with fine specimens of Alaskan king crab. *(Left: Courtesy of the Hansen Family) (Right: Courtesy of EVOL)*

Bering Sea around St. Matthew Island and the Pribilof Islands. It's so similar to red king crab that they are sometimes marketed as the same thing. The spikes on blue king crabs are black-tipped, and the meat is a deeper red than a red crab, but they are generally about the same size. They bear no resemblance to the Chesapeake blue crab. Fishing blue kings has always been my favorite, mostly because it's a short summer season way up north. It was my first crab trip ever.

Brown crab is also known as golden crab, and despite its name, it's the smallest and least desirable of the Alaskan kings. They are orange or brown in color, and a bit smaller than the reds and blues.

King crab live on the ocean floor, and do not venture to shore. Most of the year they stay several hundred feet below the surface, but in late winter they migrate to shallow water, less than 160 feet deep, to molt, mate, and hatch their young. Crabs may migrate over a hundred miles.

Molting is how they shed their hard shell and reveal a new soft shell in its place, which will harden over the course of the year. Molting allows the females to reproduce. They breed through genital ducts called gonopores on their sternums. The male grasps the female and fertilizes somewhere between 50,000 and 500,000 eggs. I'm no scientist, and to be honest, I don't totally understand how that procedure works.

The female stores the eggs under her tail flap for eleven months as she returns to deeper water. The next year, before molting, she travels to the shallows and lays her eggs. Millions and millions of embryos are released and hatched into larvae. The larvae swim and eat phytoplankton and zooplankton, or get eaten by larger creatures. They will molt four times before becoming recognizable as crab. For protection, juveniles between the age of two and four form pods made up of thousands and thousands of crabs that roll across the ocean floor in a big ball.

After four years, they may leave the pod, but they are still quite small—less than three inches across the carapace. Then they join the annual migration and spawning of adult crabs, and continue to grow toward maturity. Adults are omnivorous, and will feed on just about anything that comes their way: algae, worms, clams, mussels, sea stars, sea urchins, fungus, other crab, and of course, dead fish parts like the bait in our pots. Their main predators are large bottom-feeding fish such as halibut and cod, but if they can avoid pots and stalkers, they live ten to twenty years.

Sometimes the adults, like the juveniles, will form a pod of thousands of crab, a huge ball of biomass that rolls across the ocean floor. Just picture it—a pod that covers an area the size of several football fields, with crab ten or fifteen deep, pulsing and surging, legs and claws poking out every which way; hundreds of thousands of pounds of hard-shelled bugs swarming the ocean floor in a frenzy.

Right smack-dab in the middle of this pile is where we want our pots. Getting them there requires the elements I listed earlier: experience, science, cunning, instinct, and luck. First let's look at experience.

Typically I take the boat to places I've been before. My dad and Tormod kept detailed logs of all the crab they pulled over the years. I do the same thing—a handwritten ledger, in pencil, with the exact number of crab pulled from each pot, along with its coordinates. Probably more than a million pots pulled and logged, so we have a record of where the bugs have been. That doesn't mean they'll be in the same place next time.

We know that crab like the shallows, but we don't know exactly where they will be in the Bering Sea's hundreds of thousands of square miles. So next comes science—although an actual scientist might not call it that. We study currents, plankton, weather, water temperature, and past history. Then we make our best guess.

Then there's cunning. I always monitor the radio—if you can read the other skippers, they might lead you to the crab. There's a whole culture around radio. It's like poker. We're always lying to one another. If I'm on the crab, I don't want the rest of the fleet dropping their gear beside me. Likewise. if I'm *not* on the crab, and another skipper is *also* not on the crab, I want him to think I'm pulling them in, just to rattle him, so he'll wonder where I am, and wonder what he's doing wrong. Some boats work in teams or partners, and if one gets hot, he'll let the other know, but if he tells his partner his location, everyone else may hear. So we speak in code, disguising the number of crab and our co-ordinates in a way only our partners will understand. Before leaving port we assign a letter from the alphabet to each longitude and latitude line. Then we tell our partner we're on a bite at "Bravo Tango" and they know where to find us.

You learn to read the others. If a skipper is blabbing for a while and then goes quiet, you might guess he's on the crab. Or you guess that he's so frustrated that he doesn't want to talk to anyone. It depends on which skipper it is. So that's why it helps to have twenty-five years' experience up here. The better you know the players, the better you can tell when they're bluffing, lying, hiding, or telling the truth. Even if we're just shooting the shit, I can smell when they're lying. I know when they're hiding something. I can read through it. That's part of fishing. Every captain has to learn how to decipher each person. If a green skipper comes up here, the rest of the guys will have him bouncing around like a ping-pong ball with false leads and fakes, and he'll end up doubting himself and wondering if he's nuts.

That's where the next element comes in—instinct. Sometimes you just get a feeling that you're in the right place. Even though it might go against your previous experience or your best judgment, you follow your gut, and you find the crab. Or, you follow your gut and you get skunked. If your instincts are bad, you won't last long in this profession.

This leads to the final element: luck. Now and then you just drop your pots on a mountain of crab, thousands and thousands in that rolling biomass. You didn't earn it, or deserve it—you just got lucky. Because luck is so important, skippers become superstitious. I carry superstitions, most of which I learned from my grandfather. For one, suitcases are forbidden on the *Northwestern*. When the *Deadliest Catch* started filming onboard, the cameramen had to leave them behind and transfer their gear to duffels. No horseshoes. No bananas. In addition, I don't allow anyone to talk about having a good season before we leave.

Then there are the things I *do* want to do. I always talk to Charlie McGlashan on the way out. If the minister is around, I'll have him come aboard and say a prayer for us. Someone has to bite the head off a herring for good luck. Also, I like to have a certain type of erasers on the end of my pencils. I will not go to sea without these erasers. I wrap them with black electrical tape so they won't fall off. If I don't do it, I freak out. I also won't leave port without Post-its. As time goes on, you start to look for these silly little things, and if they're not there, you lose your mind. You'll turn the boat around and go get them. Psychotic—I know.

So that's how I decide where to drop the pots. The next question is, How will I set them? A crab boat is a floating assembly line—most efficient when all the motions are repetitive and predictable. What that means is that I drop the pots in a straight line, usually in a string of thirty. I want them evenly spaced, so that when we go to pick them up, the time required to travel between pots is the same as the time required to haul, dump, and sort a pot.

Dropping the pots in a straight line isn't as easy as it seems. Thirty-foot seas and forty-knot winds knock the boat off course. What's more, I want to set the pots so that it will be safe to pick them up. So let's say that when I'm dropping the pots, I get a forecast of a storm arriving from the north in thirty-six hours—which is when I think I'm going to be pulling them. I want to drop that string on a north-south axis, so I can jog face-first into the seas while we pull gear. If I set the string east-west, then I end up pulling the gear sideways to the storm, exposing the crew to huge waves crashing over the rail that could wash them overboard. I also have to consider the direction and speed of the tide.

People often ask me what the advantage is to having the house in the front of the boat versus the aft. There's good reasons for both. In a house-aft—a schooner—the captain gets a better view of his crew, and he doesn't take waves over the wheelhouse. When he is bucking into a storm on a schooner, he's not worried about his windows breaking. Also, a schooner usually has much better living quarters because there's simply more room.

Schooners also have two engines—twin screw—which means you can back one engine down and drive forward with the other. They're more maneuverable and you can power out of things. On a single-screw engine like mine, I have to have more momentum if I want to turn to port or starboard. The turning radius is bigger.

Schooners have a downside, though. First, if the pots are stacked on the deck, the captain can't see very well. Second, without the wheelhouse to block the seas, the deckhands are more exposed to crashing waves and more likely to be swept overboard or across the deck.

The main disadvantages of a house-forward, like the *Northwestern*, is that the captain is more likely to get creamed by a big wave, which can smash the wheelhouse windows. Sometimes the metal awning above the windows gets knocked upward like the bill of a baseball cap—what they called the Bering Sea salute. Another problem is that the skipper is constantly looking over his shoulder to see what's happening on deck. Nevertheless, I prefer a house-forward because of the better view of the seas and the protection it affords the crew. Also, between a hundred-foot schooner and a hundred-foot house-forward, the house-forward is going to fit a lot more pots. The schooner has a smaller deck, and to get the same amount of pots on it as the *Northwestern*, I would have to build a very big schooner.

When I'm in the wheelhouse setting gear, I'm doing a bunch of

Here I am in the wheelhouse. *(Courtesy of EVOL)*

things at once. First, I'm looking out the window at the seas, keeping us square to the waves. If something huge rises up, I hit a warning buzzer which gives the crew a few seconds to brace themselves before we get swamped. Then I'm steering the boat and running the throttle. I've got a panel of video monitors that show me the deck, the engine room, and the lazarette, so that I can see what's happening on the rest of the boat. If we're full with fuel and crab, and the lazarette floods, we could sink quickly, so I want to know what's happening back there. I also study my fathometer. I need to know how deep the water is, to make sure we have enough shot on each pot. As the crew drops each pot, I mark it on the computer's sea chart. When we're done, the monitor shows me exactly where the pots are.

That's what it takes to find the crab. If you do it consistently, you'll catch more than the rest of the fleet, and they'll call you a highliner.

Then you can attract the best crew, because they want to go with the one who makes the most money. You don't get to be a highliner every year. Even the best skippers with the hardest-driving crews will not become the best without a good dose of luck. Crab fishing has three factors: drive, fishermen's intelligence, and luck. With all three, you'll be a highliner. With two, you will be above average. If you only have one, you'd better find a new line of work.

Fishermen had been catching crab in Alaska for decades before my family arrived. The Japanese began canning the product at the end of the nineteenth century. Before World War II, a few Alaskan companies dabbled in the process, but well into the 1950s, the Japanese ruled the market with canned crab, especially the Geisha brand. The Japanese fleet fished Alaskan waters with tangle nets, big loosely hung webs that entangled any species with protruding spines. Once the catch was hauled on deck, they would simply cut the crab free and discard the net. Tangle nets were devastating for sea life, because even after they were discarded, they continued to catch and kill anything that wandered into them.

For Americans, crab fishing was an eccentric unprofitable enterprise. Lloyd Cannon remembers fishing as a boy in the winter of 1947 off Kodiak Island, helping his father with tangle nets.

"My father couldn't afford to cut the nets, so we sat there and tediously untied the crab," Cannon said. "If we caught twenty-five crab, it was a big day. They sold them on the street corners in Kodiak."

The crab industry as we know it today wasn't really born until the arrival of Lowell Wakefield, son of a well-known fisherman from Anacortes, Washington. Wakefield had attended the University of

Washington, and then got a master's in anthropology from Columbia, but he wanted to be a fisherman like his father.

During the war, the family began packing crab in Port Wakefield on Raspberry Island, near Kodiak. In 1945, Wakefield launched a new company called Deep Sea Trawlers, that would later become Wakefield Seafoods. He began fishing the Bering Sea in a boat called the *Deep Sea,* a brand-new, steel, East Coast–style, side-rigged trawler about 135 feet long. A trawler—also known as a dragger—drags a huge net along the ocean floor, scooping up everything in its path. In those trawl nets, everything came aboard in a big bag and was dumped on deck. There might be eight hundred males and five hundred females, They would butcher the males right there on deck, and toss the females back. Pretty quickly Wakefield was catching more king crab than he could process.

Charlie McGlashan's uncles Aleck and Steve were among the first deckhands aboard the *Deep Sea.* On its first trip to the Bering Sea, the vessel was so pummeled by waves that it took a 63-degree roll. Steve McGlashan flew from one top bunk to the other without dropping to the floor. The boat lay on its side. Another swell would have been the end of her. Instead, she righted herself. The skipper immediately dropped off the crew, steamed to Seattle, and filled the bilge with ninety tons of concrete for ballast. Returning to Alaska the *Deep Sea* became a legendary success.

One night while working in his processing plant, Lowell Wakefield had too much product to finish before the day ended. He placed the remaining crab in a cooler to store until morning. When he arrived the next day, he realized he had accidentally left it in the freezer—not the cooler. When the crab thawed the next day, it was still in great shape. He realized he'd come across a preserving method superior to canning. Wakefield got the idea to serve it in the shell.

"He was a man who had a vision that was sometimes difficult for the rest of us to see," said Dick Pace, the late president of UniSea, or Universal Seafoods, a major processor in Dutch Harbor. "He believed that king crab could be the gourmet food for the twentieth century."

Wakefield's challenge was to sell the rest of the nation on the idea that these spiny monsters made for fancy eating. An entrepreneur named Stan Tarrant set up a crab processor at King Cove. His Pacific American Fishing shipped twenty thousand cases of canned king crab in 1955, but they couldn't sell it. Finally, they offered a twenty-five-cent rebate for each label of the can. Only then did the product move.

Meanwhile, Wakefield's frozen crab gained traction. In New England, where lobster was king, the bright red king crab legs emerged as a less expensive substitute.

"The king crab industry has a terrific future," Wakefield told a reporter at the time. "Most of our traditional fisheries—salmon and halibut—are summertime fisheries. This one is a wintertime fishery, bringing in dollars to Alaska fishermen, processing workers, and businessmen at the time of the year when we need it the most."

Other fishermen took note. Ed Shields got his first inkling of the possibilities of crab fishing in 1947, when he saw Wakefield's *Deep Sea* dragging the Bering Sea. Shields and his father, J. E. Shields, were the last of a dying breed: captains who fished for cod in sail-powered wooden schooners. Not equipped with freezers, they preserved the cod in the old Norwegian style—salting and drying. That year, the younger Shields was a crewmember on his father's vessel, the *C.A. Thayer*, a 219-foot, three-masted schooner built in 1895. It was the last sail-powered ship in the fleet, and its final commercial voyage in 1950 would mark the end of an era. The elder Shields heard on the radio how much crab the *Deep Sea* was catching. He soon bought a 148-foot surplus wooden army freighter. They rigged it as a side-trawler, and called it the *Nordic Maid*.

Other fishermen started to experiment with the round pots used for Dungeness crabs in the south. But they only weighed 150 pounds each, and broke apart in the Bering Sea.

Then a fisherman from Seldovia named Joe Kurtz had an inspired idea. He came across a bunch of steel-framed surplus army cots on a local military base. "We took the springs out, and knocked the cots apart," he remembers. "It was easy to do with a heavy hammer and chisel. One whack and you cut the rivets. We stood two of them up on edge, cut the third one in two, and took the pieces we got out of that for the ends. We ended up with a pot that was sixty-three inches long—because that's what an army cot was." Thus the modern steel box-shaped crab pot was invented.

Eventually the fishermen discarded the cots and made pots from scratch, stringing chicken wire between the frame of welded steel rods. Howard Carlough remembers they could build them at a rate of four per day. Innovations were unplanned, all trial and error and happy accident. They experimented using five-gallon wooden beer kegs as buoys, but the jellyfish would climb on and swamp them. They were finally replaced with the plastic "jap buoys" in the early sixties. They built four-by-six pots, with a straight tunnel on the four-foot side, that the crabs would climb through. Then they experimented with the tunnel on the six-foot side, and that caught more crab. Then they built the things seven-by-seven, and caught more still. One day Howard Carlough saw a fisherman accidentally drop a pot from the crane onto the dock. It bent the frame, so now the crab tunnel, instead of being horizontal, pointed up at an angle. Well, they weren't going to get rid of the thing just because it was bent. They fished it. The bent pot caught three times more crab. It wasn't a fluke. It always caught three times more. "My theory is that the crab liked the bent tunnel because they could look up and see the bait," says Carlough. Soon they bent the tunnel on every pot.

Fishermen were catching crab faster than it could be processed. Cap Thomsen, a Kodiak skipper, decided to go into the processing business. He bought an old barge. He advertised in the Kodiak newspaper that he would be processing in the Aleutian Islands, and urged fishermen to come out. When the crabbers arrived, they found there was a catch. "Look, I don't have a nickel to buy a crab," Thomsen told them. "But I know where the crab are. And if you're willing to wait— I'd never processed a crab in my life—I was willing to buy your crab." In all, Thomsen invested $1610 in his Aleutian King Crab Company. Five years later, he sold it for four million.

By the mid-sixties, business was booming. The fleet's catch of king crab leapt each year: 11 million pounds in 1958, 28 million in 1960, 52 million in 1962, 86 million in 1964. In 1966, twenty years after scraping to net a few dozen crab with his father in Kodiak, Lloyd Cannon crabbed on the F/V *Juneau*; over eleven months the boat caught 5.4 million pounds and grossed $540,000. Each crew share was $54,000.

"And then the squareheads came up," remembers Howard Carlough. Perhaps the first Norwegians fishing crab in Alaska were Einar Pedersen and Ole Hendricks, who had immigrated to Ballard many years early. Now a younger wave arrived. Among the pioneers were Sam Hjelle and John Sjong and the two brothers from Karmoy, Magne and Kaare Nes. Magne came to New Bedford aboard the *Stavanger Fjord* in 1954 as a teenager. He fished a few seasons for scallops, then in 1957 drove across the country to rustle up some work in Seattle, dragging with Dan Luketa and Jackie Ray. In 1959 Ole Hendricks hired him in Alaska, and by 1962 he was fishing crab. Over the years he owned twenty-eight boats, a consistent highliner, pulling in more than fifty million pounds, probably more than any other fisherman; and that doesn't even count the 150 million pounds he threw back that were female or too small. He was also an innovator, one of the first guys to use a throwing hook rather than a pike pole to land the buoys.

Magne Nes introduced his older brother to Alaska. Kaare Ness (who changed the spelling of his last name), switched from scallops to crab. According to him, in Alaska, "there was only one speed: full speed." Kaare would eventually partner with an American named Chuck Bundrant to form Trident Seafoods. Bundrant was a lanky Tennessee boy with a southern accent who bought his first boat at age twenty-five—one of the true pioneers of the fleet. He was fearless. He and my dad were good friends and Dad used to call him "Country Charlie Pride." Chuck and Kaare's breakthrough was to build a vessel that not only caught the crab, but processed them right onboard. Four decades later, Trident is the largest American-held seafood processor in Alaska and the Northwest, with 4000 employees and over 250 million pounds of fish of many varieties processed per year. We've been delivering our crab to Chuck and his company since 1983.

Another of Trident's eventual partners was Bart Eaton, who arrived in the Kodiak canneries in the early sixties with no fishing experience. One day he was offered a job on a boat and never looked back.

"In those days you could go on a boat with your raingear and a pair of boots and actually dream that you'd own the boat someday," he said. "I was raised on a small farm. And there was no way that you could work on a tractor and ever think you were going to own the man's farm. I don't know how many enterprises there are where the laboring man, with nothing but a strong back, could dream that kind of dream."

This was the industry—and the dream—that beckoned to, and delivered, Sverre Hansen to the helm of the burning *Foremost,* that bitter dark morning of December 8.

6

DUTCH

With each muffled boom below deck, the old wooden ship shuddered. The whole thing could blow at any second, and when Captain Sverre gave the order to abandon ship, he didn't mean in five minutes or in sixty seconds. He meant now. Drop what you're doing and forget about grace under pressure and run for your goddamned life. Leif Hagen was first in the raft, flinging himself over the rail and splatting at the mouth of the tent that enclosed the rubber donut. Now he held the raft close to the rail of the *Foremost* while the others leaped aboard, a brutal dogpile on the trampoline floor. The raft felt unseaworthy. Sverre tried to push himself up from his hand and knees where he'd landed, but the squishy rubber offered no resistance. It was like trying to do pushups on a waterbed.

The raft was tethered to the windward side of the deck, and the 50-knot gales held it snug to the hot wooden hull. The fire had escaped the *Foremost*'s engine room, and a maelstrom of flame engulfed her stern. Leif cut the line—the raft was free. Yet it didn't move. The wind had pinned the raft to the starboard hull of the *Foremost*. Here they were, bobbing in a waterbed in hundreds of miles of open ocean, and the only place their waterbed wanted to be was plastered against a floating bomb. It was like riding the front bumper of a semitruck doing ninety down the freeway.

"Get the paddles!" Sverre cried, and the men flailed among the supplies on the floor of the raft. There were only two paddles, spindly two-foot-long units that looked like toys. Leif and Magne each grabbed one and leaned out the opening of the canopy, digging the blades into the frothy sea. The raft clung to the ship and Krist pushed off with his bare hands. They just needed to clear the bow of the burning boat, and then the wind would separate the two craft. If only the wind would relent for a moment they could free themselves.

Careful what you wish for. Suddenly the gale shifted direction. The *Foremost* heaved to starboard and blocked the raft's escape route. Instead of the raft merely bumping against the ship, the flaming tinderbox was now bearing down on *them*.

"Paddle!" Sverre screamed helplessly. Stuck inside the canopy without a blade of his own, there was nothing else he could do. "Dig in!"

Leif and Magne paddled like mad. The blood pounded in Leif's temples until his whole head throbbed. He threw his shoulders into the work. His hands submerged in the icy sea with each stroke as he tried to get more purchase from the dinky paddle. The *Foremost* was catching all the wind—effectively sheltering the raft so that it bobbed helplessly in the path of the ticking, floating time bomb. They couldn't paddle as fast at the wind pushed the *Foremost*. Any second it could explode, and if they were pinned to its hull at the moment of impact, the result would have been the same as if they had stayed onboard.

They cut an angle toward the bow, and then in what counted as luck for a day like this one, the wind shifted again. The *Foremost* pivoted and floated past them, a smoking hulk booming with internal explosions.

Leif and Magne stowed their paddles and caught their breath, their lungs heaving and hearts pounding. They were safe and alive, for the moment. They threw the drift anchor overboard, to slow their movement in the current, and put some distance between themselves and the *Foremost*. The burning ship drifted away. Fifty feet. Hundred feet. Two hundred feet. *Don't do anything stupid,* Sverre told himself. *Take this one minute at a time and you'll figure something out.*

Even though I'm proud of the amount of crab we've harvested and the money we've made over the years, I think my greatest achievement in life is that we've been blessed to never have had a death or permanent injury aboard the *Northwestern*. Knock on wood. The chances of getting hurt or killed are so great, that safety is a full-time job. So when I said earlier that the captain's job was to find the fish—that's actually his second responsibility. Job One is keeping the boat safe. I've done pretty well at this, but I could have done better. We've had plenty of close calls and near misses that keep us humble. If you watch the *Deadliest Catch*, you may be a bit of a disaster junkie, and if so, this may be the part of the book you've been waiting for. This is the part where I recount some of the close calls.

There are a few stories that we're taking to the grave, but one that I'm willing to write about here was when I was about twenty-seven. We were fishing opies in the middle of the winter and it was blowing 45 or 50, steady, and most of the guys that were close went into the island to anchor up. We stayed out and fished. I was grinding pretty hard, on lesser fishing than the guys who were closer to shore. So I just kept grinding even harder to make up for the small pot counts.

Deckhands Rick McLeod and Matt Bradley beneath the heavily iced wheelhouse of the *Northwestern. (Courtesy of the Hansen Family)*

Time was money, so we didn't take the time to chip the ice off the boat. The guys were fried anyway, and chipping ice required that you stop fishing, in other words, stop making money. We slept, got up, accumulated more ice. I should have ordered the crew to chip ice, but I didn't. I wanted to keep grinding.

After a couple of days, three to four feet of ice had built up over the whole boat, especially around the bow and wheelhouse. Of the fourteen windows in the wheelhouse, I was looking through the only one that was not iced. It was a heated "spinner" window that doesn't accumulate ice. It was like peering through a tunnel of ice. It got bad. The boat got so nose-heavy that she was actually going down nose first. She took a wave, and when it came up over the bow, I had water up to the windows. Even the deck was submerged; and she stayed like that. She was on her way down. All I could do was throttle out of it. I gave it all she had to starboard, but the prop was barely underwater. It wasn't doing much. You couldn't see the bow—just water up to the windows. The boat turned sideways and the next wave hit, and she keeled over. Instead of capsizing, though, the keeling allowed a lot of that water on the bow to drain out. So she righted herself, very slowly. It was eerie, like dying in slow motion. I got the boat turned around, and we went with the weather. Then we started chipping ice. It took a solid eighteen hours. We were lucky to get out of that one. After that, I never took Bering Sea icing conditions for granted again.

We've had our share of rogue waves, the bastards that come out of nowhere. One time we had come through a storm, and things looked pretty good. We had started fishing again and all the guys were on deck, hauling the first couple of pots, even though it was still blowing really bad. I had the boat at a dead stop so they could work. All of a sudden this monster wave came out of nowhere, like a building going fifty miles an hour. When it hit, I thought the windows were going to shatter. It happened in a second. I dived down and put my hand on the buzzer to give the crew a warning blast and waited for the windows to smash through.

Luckily they didn't, because I had the boat at a dead stop, which reduced the force of impact. The reason it hit us straight on is because we weren't traveling. We were facing the wave, pulling gear as safely as we could. It had just enough curl on top to where it broke over the top of the wheelhouse instead of smashing through the windows.

I was on my knees. I crawled to the door, scrambling to get up, and all I saw on deck was water and white foam from rail to rail. I figured everyone was gone. I was screaming at the top of my lungs. I thought I'd killed my family, my crew—everyone. I opened the door. There was still water coming over the top. It poured down over me, and now I was soaking wet and screaming.

Finally it started to mellow out. Then all of a sudden I saw these little heads popping up—one, two, all of them. They started laughing. They had heard the buzzer just in time and got up underneath the shelter and grabbed onto something. A rogue wave can strike just that quick. In a few minutes we went back to fishing.

Another time we were running across the Gulf of Alaska on our way to the Bering Sea; it's safer to hug the shore but we saved a day by cutting across. Out in the gulf, we got caught in a huge storm. We bucked big waves, every six or seventh of which was a rogue wave. You could time them. We'd see them coming, just curling at us, and I knew we were going to just get pummeled. I pulled the boat out of gear and jumped on the floor and hugged the base of the seat. I felt the whole boat go *boooomm,* and just shudder. I popped back up and shoved it in gear and started up. We were trying to make headway in the storm, but were absolutely getting our asses kicked. We were out in the middle of nowhere. We took turns at the wheel, and whoever was steering had to punch a button on a watch alarm every so often to make sure he was awake. If he didn't hit the button, the alarm blared.

Usually we set it for every fifteen minutes, but this storm we had to set it for five. If someone dozed off at the wheel at the wrong moment we'd be sideways to a wave and we'd all be dead. Mark Peterson remembers that we hit one of these big waves when I was lying in my bunk. They could hear me at the top of my lungs screaming, "Fuck! Shit!" They didn't know I was sound asleep. I even rolled off my bunk and crawled back in without waking up. Peterson thought that if he survived this storm, he would never come back on a boat again. We made it through, though, and the next trip he was back onboard. Fishing can become an addiction. Ask any of the guys.

A few years back I got tired of fishing opies because the numbers had been low. I decided to fish tanner crab, a slightly different species of snow crab. It was open season on them, the price was better, and just a few boats were fishing them. We left the western opilio grounds and traveled four hundred miles to the eastern Bering Sea to tanner grounds, past the Pribilofs. It was a risky move for an inexperienced kid—it was very far away, would cost a fortune to get there, and we didn't know if we'd catch anything once we arrived. I raced to get there to save time and money. The boat had a full load of gear, which makes it very heavy with a slow roll. Suddenly the boat took a rogue wave. I just had too much speed. She came down, *bam!* Then she rolled way over to one side. The shelter wall on the portside with the crane caved in. It smashed the railing and ruptured the fuel vents, and we couldn't fix it with the welder for fear of an explosion. In addition, our empty fuel tanks were taking on seawater. We kept her upright by transferring fuel from one tank to another, to compensate for the list, and we limped back to harbor for repairs. The lesson I learned on that trip was to watch my speed no matter how much of a hurry I was in. Mother Nature always wins.

On another trip we took a rogue wave from the stern. It came up and over the deck, over the top of the wheelhouse. So much water came down that it rushed through the air vents on top of the house and blew out the ceiling of the bathroom. We hosed it down with fresh water and kept going. We fixed it at the end of the season.

The same thing happened another time on the other side. We were just about to head in, and these waves suddenly came at us. I turned the boat too hard, gave it too much throttle. She came up and just dove down into this thing. The next thing I knew she was on her side, buried in the nose. All the water on the deck poured in the galley window, because it wasn't dogged shut. I went hard over, and turned it back. I got a feeling in my groin, like the funny feeling you get on a roller coaster when you're stuck in place, helpless, and screaming. Finally, she began to slowly come back up, gasping for air as she righted herself.

Another time we were just about full with only a few more pots to pull. We would have been one of the first boats in. We'd started way down south and found a good spot. It was the first trip so we had a lot of fuel onboard, which makes a boat heavy. We also had all three tanks of crab pumped down, full of water and crab, which made us heavier still. In that condition a boat doesn't respond very well. It's like a rock in the water, a submarine with a really low center of gravity.

It was a really rough night so we stopped fishing for a while. After the storm ended, it was flat calm. We couldn't see a single whitecap in the darkness. We decided to start fishing again to top off the tanks. I was real happy about the catch and to be heading to port so soon. The next morning, I played some music in the wheelhouse to celebrate.

Brad Parker was the engineer. He was a super good worker—one of the best. He was one of five brothers, all of whom had fished up there. His father, Dick Parker, had been the *Northwestern*'s engineer before him.

Suddenly, out of the corner of my eye, I saw it coming out of nowhere: a rogue, just a lump, no crest. Then it began to break, a mountain of whitewater. I quickly hit the buzzer, but I was a half second too late. The water hit us at a 60-degree angle. Brad was right at the rail. It picked him up and smashed him into the coiler. Then it picked him up a second time and threw him all the way to the other side of the deck. We thought he was dead and ran over to him. He started moving a little bit. He was very smart—he didn't move too quickly. We stood around him, not knowing what to do. We didn't want to touch him.

Brad managed to crawl inside the galley on all fours. We helped him into his bunk and lashed him down so he wouldn't get tossed around. We stopped fishing immediately, cut the trip short, and headed to town. We went in to port and got him to a doctor, and he was sent home. It wasn't until he reached the hospital in Seattle that he learned he had chipped his spine.

To this day, I wonder if I hadn't have been listening to music, would he have gotten hurt? Would I have paid that half second more attention? After that day, I *never* again listened to music in the wheelhouse while fishing. Ever.

Back to the saga of Sverre. When we saw him last, the year was 1964 and he'd just married my mother and been offered his first job on an Alaskan crab boat. These were two of the most important events in his life, setting into motion all that came next. They sure determined my own life's path.

When I hear the stories about those early years in Dutch Harbor, I'm just blown away at those men's courage. Of course it takes courage to do what my brothers and I do now: the monster waves and killer storms you see on TV are all real. Even so, when my brothers and I fly up to Alaska, at least we know what we're getting ourselves into. We know the risks, and we know that if we pull it off we're going to make a buck. When my father and my uncle first went to Alaska, they had no idea what to expect. It was the frontier. There was a good chance they wouldn't return. Also, there was no guarantee they'd make a cent. It really puts what we do into perspective.

Karl Johan was actually the first of the Hansen brothers to fish for Alaska crab. After two years on a Seattle dragger, he went up to Kodiak Island in the summer of 1963. The guy he fished with, an American skipper from Tacoma, was a hell of a good fisherman, but the year before, he'd had a tragedy. He had moved his family from Tacoma up to

Kodiak, intending to live there permanently. He sailed a forty-five-foot boat called the *Tahiti*.

So he brought his wife, son, and crew up to Kodiak. They stayed there a few hours. It was blowing like hell. People told him not to go around the point to Port Wakefield. He left anyway. There was no ballast in the boat—just the furniture he was moving to his new house. He came around the point and the boat tipped over. He got his son on his back and swam him to shore and put him on a rock. He went back out to the boat for his wife, but she had already drowned. He couldn't find her. He swam back to the beach, but his son was gone, washed away. He had lost them both.

Karl fished with him the very next year, and the skipper was a bit nervous. After six months, Karl flew home to Seattle without enough money for a taxi ride home. He doubted that crab fishing was the wave of the future.

Sverre, however, liked crab fishing. I've seen pictures of him working with Howard Carlough on the *Western Flyer* in 1964, a plaid cowboy shirt with the sleeves rolled up, an engineer's cap tipped to the side, and a smirk on his face. He looks like a cross between Edgar and Norman and me. He and his crew fished up and down the chain from Adak to Kodiak. Back then the fishermen were responsible for unloading their own crab. One day the *Western Flyer* steamed into Akutan to unload onto the *Deep Sea*, which by then had stopped dragging and was strictly a floating processor. Sverre's old friend Charlie McGlashan was out on the city dock.

"Wanna unload crab?" Sverre called.

"How much you pay?" Charlie asked.

"Twenty bucks." That was a lot of money in 1964.

So the crew tied up and walked up to the Roadhouse to have a few cans of beer while the kid got to work hauling crab from the tanks. "I unloaded six thousand crab all by myself," Charlie remembers. There were no brailers—the crane-powered huge baskets like you see today—just big steel buckets. Once loaded, they were lifted by crane aboard the *Deep Sea*, dropped into a live tank from which they were picked, butchered, and frozen.

They also fished right outside the mouth of Dutch Harbor. Without sodium lights, they were unable to fish at night. The crew was up each morning at five, on the water all day, and then back at six. I think the Old Man loved those days, and later, when the competition and technology allowed—or forced—crabbers to work day and night without rest, Sverre missed the simplicity of the old days. It must have been cool back then, without all the contracts, schedules, quotas, laws, and regulations that we have now. But would I go back to those days? Nah. I'm willing to put up with all the B.S. If you know how to work the system, you can make more money at it. We don't mind working the hours we do. Hard work is rewarding. We're willing to go forty-eight or seventy-two hours without sleep if that's what it takes.

My mother, Snefryd, didn't know anyone in Seattle yet, and didn't speak English, so she and Howard Carlough's wife flew to Dutch Harbor to spend the summer with their husbands. After staying three weeks with the pilot of the airplane, Sverre and his young bride moved to a little wood-slat shack, a duplex in a slum of three identical buildings. The refrigerator was a wooden box propped against the outside of the window. Mother would open the window to get at her milk, cheese, and sausages. The only place to buy food was the store in Carl's Hotel. Some days Snefryd went out on the boat with them, but sometimes she got seasick and stayed home. It got pretty boring at home. In addition to full-share deckhand, Sverre was also the cook. There was an enormous stove aboard the *Western Flyer*, and each night when he got home, he baked fresh bread from scratch, which they would eat for breakfast, and then use for sandwiches on the boat.

In 1964, Dutch Harbor didn't have any of the large canneries and hotels. Alaska had only been a state for five years and the crab industry was new. When Howard Carlough first arrived, the entire fleet consisted of five boats—two steel and three wooden. They had sixteen-mile radars, but half the time they didn't work. You had to know how to fix everything on your boat, because you couldn't call a plumber or a mechanic. There was no dry dock or repair shop. They had to make do with what they had. The Old Man told me about a boat that lost a bearing out in Adak. They were totally disabled. Two other boats arrived to help and anchored on either side to form a makeshift dry dock. They hoisted the stern of the damaged boat entirely above the surface of the water to expose the shaft so they could work on it. However, the one replacement they possessed wouldn't fit on the shaft. In an inspired moment of mechanical ingenuity, they boiled the bearing in oil for three days to expand it. It worked and it finally slipped on the shaft. The boat was able to continue. It's this type of camaraderie and innovation that impresses me to this day.

Fisherman could figure out a way to do just about anything. Probably the most hard-core thing I've heard about those early days was told to me by Carlough. He remembers that if a crewmen had a toothache, they just squeezed a drop of battery acid into the cavity to seal it off. That's right, battery acid.

Before I continue with Sverre's saga, I need to tell you about Dutch Harbor. It is, after all, the place where most of the story went down. Now that it's become the "set" for the *Deadliest Catch*, it's developed a sort of mystique. Some fans of the show have been known to fly up there just to see the place. If you'd have told me ten years ago

Gearing up in Dutch Harbor. *(Courtesy of EVOL)*

that tourists would be spending over a grand on airfare to go have a look at this windblown, abandoned-rusting-warehouse-not-a-single-tree-grows-there company town, I'd have laughed.

It takes a special type of person to live in Dutch. The locals are great, one of a kind, and they've really improved the town in the past few years. For most, though, the place is a bit too rugged. Dutch is not where fishermen go for pleasure. It's our place of work. That said, I've spent nearly half my life there, so I've grown to love it.

Despite all the advances in technology since my dad's first days in Dutch Harbor, the place is still far away from just about anywhere, and getting there—even by plane—can be an adventure. The airstrip in Dutch is precarious—like a toothpick laid down in a rockpile—so only small planes can land there. The clouds are often too low and thick for the pilot to see the runway, but since the weather changes so quickly, the only way to know whether or not it's safe to land is to fly out there and see. The skies are usually clearest at first light, so those planes must be airborne well before dawn to arrive in Dutch for sunrise.

What does all this mean for fishermen trying to fly in in the dead of winter for opilio season? We have to congregate at the Anchorage airport four hours before daylight to board an icy twin-prop for a bumpy three-hour flight, with no guarantee of actually landing. Sometimes fishermen sleep on the airport floor the night before rather than pay for an overpriced hotel for a few hours' rest.

My trip last winter is pretty typical of how it goes. When the attendants open the doors, hundreds of fishermen and cannery workers crowd the desk hoping for a seat. All of the previous day's six flights have been canceled because of weather, so now hundreds of passengers are vying for the available tickets—all of which have been booked months in advance. PenAir—the only airline that serves Dutch Harbor—is trying to run six flights a day, and every slot on the thirty-seat planes for the next ten days is accounted for. Crammed together beneath the bright glare of the fluorescent ceiling lights, tired and disheveled, is a global mix of ethnicities—white, Filipino, Mexican, Vietnamese, and African.

Finally the attendants start to call out the names for the flight. The lady working the desk can't pronounce most of them, because they are not English. When they call my name and I check in, they ask my body weight. It's always a bit nerve-racking to think that the plane is so small that a few extra pounds are going to matter. They open the gate to the tarmac and the cold air rushes in. It's seven below zero, pitch-black outside, and I lean into the wind and trudge toward the plane. Snow crunches under my feet. Subzero winds whip dry snow in my face, freezing my nostrils. I climb aboard the plane. If there are assigned seats on these flights, no one pays any mind. They sit wherever they can. We're excited that we've been chosen to fly—and don't have to wait all day in the airport—and people are rowdy on the plane, yelling and heckling.

The plane has been sitting out in the subzero weather overnight, and it's cold inside. People throw their hoods over their heads and scour the overhead bins for blankets. One guy puts on a snowmobile suit. This is one of the world's few airlines where the stewardess wears a knee-length parka as she works. She hands us each a pair of ear plugs. The engines haul. We are glad to be on our way.

We take off in the dark and after an hour and a half land in the dark in King Salmon to refuel. These small planes can't make it all the way to Dutch Harbor on a single tank. We aren't allowed off the plane. We take off again, and when I awaken we are in Cold Bay. Now they let us off the plane. It is still dark at 9:00 A.M., and we have to wait for daylight to fly into Dutch.

Three planeloads of people shiver in the waiting room at Cold Bay. That's ninety people in a room made for thirty. All the benches are taken and everyone else is huddled against the walls. Half the people are lined up for the bathroom. I go outside for a smoke. The wind is whipping snowflakes in the darkness.

We take off again at 9:30, but daylight has not improved visibility. We fly through fog so thick you can't see past the tip of the wing. We need three miles of visibility to touch down in Dutch, but the weather report is saying we have only one mile. The pilot circles high above the harbor. Patches of light break though the fog bank, but it isn't enough, and after an hour of circling we are low on fuel. We have to return to Cold Bay. Now people are getting hungry and cranky. The stewardess announces that there are no more snacks on the plane and suggests the store across the street from the airstrip. A few people stomp through the blinding snow for a foam cup of noodles or a microwave burrito.

We mill around the Cold Bay airstrip. The floor is slushy with muddy boot prints. Finally they call us back to our plane, but once we're seated the attendant announces we're not going to Dutch, but back to Anchorage. We spend the entire day on the plane and wind up exactly where we started.

I spend the next three nights marooned at a hotel in Anchorage. Now that my flight has been canceled, I'll have to go standby. Each day I put my name on the list, but each day the planes are full, or they turn back from the weather, or both. On the third afternoon they promise us a seat on the last flight of the day. They say that if they can take off by 3:10 P.M., they will arrive in time to land in daylight. When the time comes, they cancel the flight. It's getting desperate now. The season starts in just a few days.

Finally on Day Four, we fly on clear skies out over the frozen Aleutian Island chain and approach Dutch about 10:00 A.M., just after first light. Mist is rising off Iliuliuk Bay as the pilot floats in from the north. Past the Spit, I can see the *Northwestern* at the dock. We bank off the steep snowy mountains of Unalaska Island, make a sharp right-hand turn, and line up over the runway. We touch down, bounce once, and immediately the pilot brakes so fast that I "heave" against the seat belt. The plane roars to a stop. We've made it.

Dutch Harbor sits in a most remote and unlikely place. The Bering Sea is 890,000 square miles, the world's eighth largest body of water. The southwestern half of the sea is a basin more than twelve thousand feet deep, not much use for crab fishermen. It's home to the underwater Zhemchug Canyon, the largest canyon in the world. More than seven thousand five hundred feet from top to bottom, it dwarfs the Grand Canyon.

The northwest side of the Bering Sea sits on the continental shelf. Here the water is relatively shallow and rich with nutrients and sea life. This shelf break, also known as the "greenbelt," extends from Bristol Bay and the Aleutians in the south, all the way north to the Bering Strait, the narrow channel where Alaska and Russia almost meet. Sea life loves these shallows, where the water is usually less than three hundred feet deep.

As many as 450 species of fish and invertebrates populate these waters, as well as 50 species of seabirds and 25 species of marine mammals including sea lions, seals, and otters. It's fertile ground for crab, as well as pollock, halibut, salmon, and many other types of fish. The Bering Sea provides 3 percent of the world's total seafood catch.

Most life springs up along the Aleutian Islands, an underwater mountain range stretching from Alaska to Russia. With more than two hundred named islands spread across one thousand one hundred miles, it's the world's largest archipelago of small islands. Twenty-seven active volcanoes and another thirty dormant ones rise out of the chain. Shishaldin Volcano on Unimak Island juts 9,372 feet above the sea—and more than six miles above the ocean floor.

The region's weather is dismal, with two hundred days of rain and snow, about fifty inches of precipitation per year, and an average of 90 percent cloud cover. Winds are strong. Winters are long. Trees do not grow unless planted and maintained by humans. Birds love it, though. The islands are home to about 250 species including puffins, snowy owls, bald eagles, and parakeets. There are more nesting seabirds in the Aleutians than in the rest of the United States combined.

The chain is largely unpopulated. During the Cold War, the United States exploded three atomic bombs on Amchitka Island. The handful of villages—Akutan, Atka, Nikolski, and False Pass—have fewer than a hundred year-round residents each. Adak housed more than five thousand in its naval base—now it's down to three hundred. Unalaska and Dutch Harbor are by far the biggest, with a combined population approaching five thousand.

With its protected harbor and freshwater stream, Unalaska has been a trading post for centuries. Eight hundred miles southeast of Russia and eight hundred southwest of what is now Anchorage, it bridged the old and new worlds. Its original inhabitants were Aleuts, who survived almost entirely from fishing, as the harsh winters make farming next to impossible. All that grows in this treeless land is wild grass and berries. The inhabitants called their home Ounahashka, which means "near the peninsula."

The Aleutian Islands have as rich a maritime history as anywhere, and it's an honor to be part of that tradition. In 1741, Russian explorers—the first Europeans to see Alaska—ventured east from Kamchatka. Captains Vitus Bering and Aleksei Chirikov crossed the treacherous sea in their boats, *Saint Peter* and *Saint Paul.* When they reached North America, the boats were separated by a storm. Chirikov followed the mainland to explore southeastern Alaska.

The Danish-born Bering was not so lucky. He fell ill, and steered the *Saint Peter* west along the Aleutian Islands. He hoped to find a route back to Russia, but was shipwrecked on a bleak island off the coast of Kamchatka. Bering and twenty-eight of his men died of scurvy. The remaining forty-six men salvaged the timbers from the wreck, built a forty-foot boat, and sailed home alive. Today, the Bering Sea and Bering Island take their names from the man who lost his life exploring them.

Bering and his crew did not find Unalaska. That honor fell to Russian traders who arrived eighteen years later, in 1759. Things didn't go so well with the thousand or so Aleuts they found living there. Fighting between the two groups continued for years. Many people were killed. Many boats were sunk. Eventually the Russians overpowered the natives, and within ten years they'd established the harbor as a trading outpost.

Unalaska became a stopover for traders and travelers, including Captain James Cook. One of the most intrepid explorers of all time, the British captain is famous for his pioneer journeys to Hawaii, Australia, and New Zealand. In 1778, on the third of his major expeditions, Cook sailed the entire northwest coast of North America on the *Resolution* and the *Discovery,* beginning in Vancouver, which they named after his crewman George Vancouver. They ventured north, looking for a Northwest Passage back to England through the Arctic Circle. Cook sailed through Sitka Sound, sighted Mt. Elias from the Gulf of Alaska, and gave Prince William Sound its name. Cook and his men spent two weeks exploring Cook Inlet, thinking it would lead to the passage, only to be frustrated to find that it dead-ended at river mouths. Legend says that Turnagain Arm, a narrow bay within the inlet, gained its name because at each dead end, the captain ordered the crew to "turn again."

Cook then sailed east along the Aleutian chain. One day in June, while sailing fast in thick fog, the ship almost ran aground on a rocky island. When a watchman heard the sounds of breaking waves, he averted a wreck. Unaware that the island already had a name—Unalaska—Cook proposed to call it Providence Island, a name that never stuck.

He then set a course north across the Bering Sea and threaded the Bering Strait. Abandoning the search for a Northwest Passage, he hoped to instead follow the Russians' Northeast Passage, along the northern shore on Siberia, back to St. Petersburg, and from there to England. But in mid-August the arctic ice pack descended and blocked his route.

The two weeks wasted in Cook Inlet would prove to be Cook's undoing. He became ill, frustrated—perhaps even crazy—and began to treat his crew cruelly, for the first time in his career. One night he ordered dozens of walruses slaughtered and forced his men to eat the beasts, even though they found the meat and oil disgusting. Three days later they reached the shores of Siberia, where the ice was rapidly descending. Cook retreated south. A month later, he dropped anchor off Unalaska, where he stayed for three weeks to repair the leaks on the *Resolution*, meet with Russian fur traders and the Aleuts they employed, and pore over navigational charts with the Russian boss.

Stymied by the cold winter, Cook set sail for Hawaii. He planned to return the next year to continue his explorations of the northern region, but it wasn't to be. That winter he was killed by Hawaiians at Kealakekua Bay, his body torn apart and burned. The search for a Northwest Passage would go on without him.

A decade later, the Spanish got into the action. In the summer of 1788, the *Princesa* and the *San Carlos* sailed into Dutch Harbor looking to claim territory for the king. The captains, Gonzalo Lopez de Haro and Esteben Jose Martinez, had sailed all the way from California, and after five months at sea, had gotten sick of each other. Lopez de Haro had even tried to ditch his companion. Martinez arrived in Dutch Harbor first, and started drinking with the Russian boss. By the time Lopez de Haro anchored, Martinez was drunk and cocky, the way men often get in Dutch Harbor. He threatened to fire his partner, but was talked out of it.

Martinez wasn't particularly impressed by the settlement of Aleuts and Russians. He guessed the town didn't have the clout to resist the power of the Spanish crown. On August 5, 1788, in a quiet ceremony he claimed Unalaska for Spain and renamed it Puerto de Dona Marie Luisa Teresa. Two weeks later the Spaniards departed their new colony. There is no record of whether the locals cared—or even noticed—that they were now subjects of King Carlos III. At any rate, the Spanish never returned, and Unalaska remained a Russian territory for almost another century. The Russian Orthodox church consecrated in 1826 still stands today, just down the street from the Elbow Room.

Then, in 1867, the upstart United States of America purchased all of Alaska for $7.2 million. At roughly two cents an acre, I estimate that Uncle Sam picked up Dutch Harbor for about fifteen bucks. Arctic foxes were introduced to many islands to be farmed. When the Alaska Gold Rush began around the turn of the century, Dutch Harbor was a coaling station for the steamships heading north for gold.

In 1940, the United States opened military bases at Dutch Harbor, fortifying for war with Japan. More than fifteen thousand men were stationed there, and the navy built a secret submarine base. Japan attacked Alaska in a little-known battle on June 3, 1942, largely to divert America's attention from the huge naval engagement at Midway. Japanese planes dropped bombs on Fort Mears, killing dozens of men. The fighting raged for days. The Japanese had believed that the nearest U.S. airbase was five hundred miles away in Kodiak, but U.S planes responded quickly from a secret airfield disguised as a cannery on nearby Umnak Island. According to the National Park Service, the brutal conditions of the Bering Sea proved almost as deadly an opponent as either army, "Soldiers shot against their own in the fog; unable to penetrate fog and clouds, ships were thrown against rocks and sunk in heavy seas; pilots met the sides of mountains in low overcast skies, or flew off course never to be seen again."

A few planes were destroyed on either side, and the battle proved inconclusive. Both sides quickly determined that the Aleutian Islands were too miserable a place to conduct a war, and the naval action moved to the South Pacific. The military presence remained in Dutch

Harbor for decades, though, giving the town its tough character. The Elbow Room opened at the end of the war, and was a hangout for soldiers before it became a fishermen's bar. There was a law that you could only have one drink per visit, so there was a revolving door. The military guys ordered a drink, drank it, went outside and got back in line.

If Dutch Harbor was too dangerous a place to run a warship, no one said so to the crab fisherman. After Alaska was admitted to the union in 1959 as the forty-ninth state, the fishermen arrived in droves. The fishing grounds of the Bering Sea were a true frontier, generally considered too dangerous for the wooden vessels that comprised most of the fleet that had for decades fished the calmer waters of Bristol Bay and Puget Sound.

Crab fishing presented two dangers to the fleet of old boats. First were the pots. To move them from one place to another required stacking them on the deck. This made for a top-heavy craft that was quick to capsize. Since crab season was during Alaska's winter, these stacks of pots accumulated ice on every surface, adding tons of additional weight and instability. The second danger was in the tanks. Unlike most fish, which are gutted, frozen, and iced onboard, crab must be delivered alive to the processors. So the holds of the old boats were converted to water-holding tanks in which the crabs could survive a week or so before delivery. Whereas iced dead fish provided stable ballast, tanks of live crab were the opposite: a loose sloshing mass that tipped with the hull. When the boat listed to one side, the sloshing of the tank actually worsened the roll instead of correcting it. As the industry grew, the numbers of boats that capsized and men lost at sea grew at a frightening rate, but that didn't stop anyone.

———

That brings us back to the saga of my dad. That summer of 1964, Sverre Hansen joined the ranks of men migrating with the crab across Alaska each year. They started in Kodiak in June and moved up the line to Chiknik, Sand Point, Dutch Harbor, and then Adak until March or April. Then back to Seattle to repair. In June it started again. (One summer in Dutch Harbor was enough for my mother—she never went back.)

The industry boomed. According to the National Marine Fishery Service, the harvest for all king crab fisheries jumped from 86 million pounds in 1964, to 131 million in 1965, to 159 million in 1966, a $15 million catch representing a twenty-fold increase since 1956. The season was limited, not by government quota, but by how much cash a processor had, how much inventory he could hold, and how fast he could sell it. "When he got enough as he thought he could sell, everyone stopped fishing, and he went down to try to peddle his product," remembers Lloyd Cannon. "It wasn't the season that stopped the fishing, it was the processors' ability to sell the product."

The other thing that could stop the fishing was shipwrecks. With those old wooden boats bobbing on the Alaskan waters, you never knew if you'd make it home alive. Here's a story that shows just how risky it was in those early days while they were still trying to fortify wooden boats for the Bering Sea.

Sverre, Howard Carlough, and the crew were crossing the Gulf in the *Western Flyer*, which Luketa had just retrofitted with the latest innovation—a molded aluminum crab tank. What its designers hadn't taken into account was the amount a wooden boat would bend and flex. Now, hundreds of miles from shore, the new tank, which was filled with water for ballast, cracked and began to flood the engine room with thousands of gallons.

They were in rolling seas, nothing too bad, but Carlough couldn't get the boat to turn. There was an observer aboard, a Chilean who was interested in learning crab fishing and taking what he learned back to his country. He didn't speak much English. Carlough reported his dilemma to the Coast Guard. He couldn't pump out the engine room fast enough. If they didn't get help, the boat would sink. A Coast Guard airplane launched immediately, equipped to deliver a bilge pump. Carlough ordered the crew to inflate the life raft.

"If we get on that thing," he said, "you'll have to hold it tightly to the deck."

Someone asked why, and the captain made a startling confession.

"I can't swim."

The Chilean observer understood well enough. He patted Carlough on the shoulder.

"Not worry," he said, making dog-paddle motions with his hands. "I swim good. Swim very good."

Finally the plane arrived. The men let out a whoop. It circled overhead and dropped its cargo. The package landed on deck, right on the money. Carlough put the crew to work pumping the engine room while he returned to the wheelhouse to see if he could get the steering to work. He finally felt relieved, like they might survive this all right.

Then he smelled something burning. Hot greasy smoke wafted up from the galley. He leapt down the stairs in a panic, and rushed into the galley.

There was Sverre, in his cowboy shirt and jeans, frying something on the stove in a big cloud of smoke. Then he noticed that it didn't smell like fire, it smelled like food. Sverre turned with a big smile, showing off the five thick sirloins sizzling in iron pans.

"What the hell are you doing?" asked Carlough.

"If we gonna go in the water," said Sverre, "we gonna go in with a full stomach!"

That's the kind of guy my dad was. No matter how serious things became, no matter how scary or grim it looked, he could always find a reason to laugh about it. That's something my brothers and I have learned from him.

Not all crab fishermen are lifers. Some work a few years and then move on. A lot of people who've seen the show wonder what a deckhand would do after crab fishing? They think all fishermen are adrenaline junkies who become skydivers or tornado chasers. Surprisingly, a lot of fishermen go on to have pretty normal lives.

Take for example Chris Aris, the greenhorn on our brown-crab-from-hell season. When he started with us he was just out of high school, lived with his parents, and hadn't a clue about what to do with his future. Suddenly, he started making a hundred grand or more per year—and that was in the eighties—but he didn't blow it all. Aris speaks, "A lot of guys don't have the discipline to save it or invest it. It seems like a lot of money, and they think, *I'm going to go get this, I can get that, I can have all the fun.* But before I started making big money, I already I had my hard-core partying out of my system—tried and done, thank you very much."

Instead, he bought his first house when he was twenty-two. "I bought a house before some guys who'd been fishing longer than me. So it depends on what you do with it. After meeting Mark Peterson and seeing what he'd done, that was more the guy I wanted to model myself after."

After five years he'd made enough money. He went to college for computer programming, and has had a job in information technology ever since. He's married and has a son. Ever since the show, his mild-mannered office mates have been shocked to learn about his past. "People are like, 'You did what?'" says Chris. "I'll say I was on there for four-and-a-half years. 'You're kidding me?'"

Aris doesn't really miss crab fishing. "When I watch the show, I re-live enough of it through that. There's parts of fishing that I always enjoyed. Coming into town with a lot of crab. I miss the home pack— all the stuff I'd bring home to stuff the freezer." The main thing he misses, however, is being part of a crew.

"You work for guys for a long time, and you're living with them in close quarters. They're really your brothers, your family," he says. "You respect them that way, you treat them that way. It's a real cama-raderie. That's the part I miss. I don't see that so much in my other jobs. The people I work with now are good to work with, but there's a part of being with somebody all the time and—apart from the times when you wanted to strangle them—there's something really cool about all that. Being part of your brothers, your family, your team."

Another deckhand from our boat who successfully made the tran-sition out of fishing is Mark Peterson. After a few years on deck, he got off the boat to start a construction company. He did that with a partner for a few years, then got tired of it. And he thought, *Well, the only other thing I know is fishing,* so he went back to it. He did another season or two on the *Northwestern,* then got a job dragging on the *American Beauty.* "I was engineer on the boat, which was a great job, but usually the next job is captain or mate, and the guy who ran the boat wasn't going anywhere. I didn't want to go to another boat because I think once you've worked on the *American Beauty,* there's nowhere else to go. You've made it. Nothing against the *Northwestern*—pollock is a totally different fishery than crab, and in those years, pollock had better working conditions and more money. I thought I had run my course."

For years, instead of blowing his earnings he'd been investing in real estate, fixing up rental properties and getting another stream of income. Now he started another construction company, this time on his own, and has been doing it ever since. Seven years ago, he took the test and got hired at the fire department. Now he's a paramedic. He still runs his construction company, and he lives in a beautiful house that he built in the neighborhood, has a wife and two kids. It's an impressive story for the kid who moved here from that poor part of Massachusetts and couldn't believe that all the kids on the block had their own cars. But the Bering Sea is still in his blood.

"I still have dreams about being up there," he says. "I dream all the time that we're on the *Northwestern* and we're going through Ballard on the streets, on the boat. Strangest dream. And I've talked to some other guys and they've had that dream, too. It's a bizarre thing. I think it's just because we spent so much time up there, and so much stuff happened, that I don't think it really sinks in until long after you're gone."

During the brief home stays in Seattle, life was good for Sverre and the rest of the Karmoyverians. After the first influx of Scandinavians at the beginning of the century, Ballard was enjoying a second heyday, as the new wave mingled with the established *norskamerikaner.* My dad's timing couldn't have been better. He arrived just as the previous generation was gaining some power, so that by the time my brothers and I were born, my dad's generation of Norwegians had fully assimilated, bought the big house in the suburbs with the two-car garage, and made a better life for their kids.

Here's what I mean when I say the Norse had become respectable in Washington. For three decades, beginning in 1953, both of the state's U.S. Senators, Warren Magnuson and Henry Jackson, were sons of Norwegian immigrants. From 1949 to 1957, even the governor, Arthur Langlie, was a squarehead. In 1962, Seattle's city fathers erected the statue of Leif Erikson in Ballard's Shilshole Marina. In 1968, the U.S. Postal Service honored Erikson by putting him on a postage stamp, finally recognizing the importance of his travels to the New World. That same year, King Olav of Norway visited Seattle, held a ceremony at the Erikson statue, then took a cruise on a crab boat. He visited the Norse Home, a retirement residence largely for immigrants, and listened to a concert of the Ballard High School band. A few years later, when Seattle became the sister city of Bergen, Norway, the king returned and planted fir trees at a small park in the heart of Ballard, christening the place "Bergen Square."

The streets of Ballard rang with the musical rhythms of spoken Norwegian. The young blond moms pushed their babies in buggies up and down the streets. Inky eventually sold the Ballard Tavern (he got a better job running a salmon cannery), and Malmen's Fine Foods was replaced by Hattie's Hat, but new Scandinavian businesses were still thriving. The Norsemen Cafeteria on Market Street served pancakes, coffee, and smorgasbord. Just across the street was Johnsen's Scandinavian Foods, where you went for pickled herring and lutefisk. There was also the Scandinavian Bakery on Fifteenth Avenue, just next door to the Norwegian Sausage Company.

On a clear spring day, Sverre drove down Market Street. Across the sound, the Olympic Mountains leapt skyward and shimmered white with snow. He was on his way to Johnsen's Scandinavian Foods to stock up on old-country delicacies for the season in Alaska. He wanted to eat Norwegian food, but the shop wasn't cheap. You couldn't go in hungry or you'd buy too much.

First came the chocolate: *melkesjokalade med hasselnotter*, and a dozen other imported bars. There were fresh-baked cookies, *spritz* and *krumkake*, and walls of sweets: preserves of gooseberries, lingonberries, raspberries, cloudberries, and black currents. Sverre didn't have much of a sweet tooth. After grabbing a few chocolate bars he went straight for the butcher's case, where he found the sausages of his youth, the same stuff he used to make at Vassvaag's shop in Karmoy. *Fare polse* is a smoked and dried lamb sausage, and *rulle polse* is a lamb roll that looks like a brick of Spam. Then there was pork sausage in wieners—*wiener polse*—or in a roll called *sylta*. Sverre suspected that he could make sausage as good if not better than Johnsen's, but he didn't have the time to do that these days, so he asked Mrs. Olsen to wrap up a few pounds of each. Then he would need a tub of Swedish spicy mustard, and perhaps just a couple of meatballs and fishcakes to snack on while he shopped.

With the meats taken care of, it was time to turn his attention to the fish he loved: smoked cod, salt cod, and herring, glorious herring! The silvery little fish came pickled in six varieties: in the traditional marinade of salt, vinegar, carrots, and onions; in sour cream; in mustard; in dill; in tomato; and last of all, in the caraway liqueur, aquavit. Sometimes he'd just get the dried herring and chew it like jerky. He bought tubes of caviar and jars of whole anchovies and cans of *fiskeboller*—fish balls in brine. For sandwiches he bought loaves of imported flatbreads and black rye crisp breads wrapped up in paper. He bought red cabbage and pickles, and tubs of Danish butter.

The Hansens celebrated Christmas at home in 1965, Sverre and Karl dressed up in black slacks and pressed white shirts and ties. Sverre was making good money. He and Snefryd had big plans for the new year. He'd lined up a new job with John Johannessen aboard a crab-bing vessel—the *Foremost*. The next month he would take the oath required to become a U.S. citizen, putting him one step closer to be-coming a skipper.

They were leaving the Ballard apartment and buying a home one hundred streets north, in the leafy suburbs. It was a big house sur-rounded by trees, with a basement and a garage. Later in the year, Sverre and Snefryd would hold a wedding there, when Karl brought his fiancée, Else, over from Karmoy. The brothers fixed up the un-finished basement for the party, wrapped a big wooden plank in alumi-num foil to make a bar. It was a potluck, and the Karmoy fishermen and their wives brought cakes, sandwiches, and booze. Else hardly knew the guests at her own wedding, but at least they were from home. They stayed up late, drinking, playing guitar, and singing.

Most important, the new house would have a lawn to play on. They would need that lawn, because that Christmas Snefryd was pregnant. By April the couple would have their first son. Lucky me.

7

A FLEET IS BORN

Sverre saw it before he heard it. Bobbing in the raft, he looked across the whitecaps and saw something strange. The *Foremost* seemed to leap out of the water.

By now the unanchored wooden ship had drifted a mile from Captain Sverre and his exhausted crew huddled in the raft. Fifteen minutes had passed. The wind howled and the men stopped panting. They had drifted past the mouth of Akutan Bay and were moving east with the current. They could see the snow-covered, seven-hundred-foot sea cliffs along Billings Head, on the northern tip of Akun Island. They would continue to drift toward Unimak Pass, where the seas would get rough. There were no towns or villages anywhere near them. Suddenly it seemed quiet.

Like a hallucination, Sverre saw a thin strip of daylight between the keel and the sea—then a flash, followed by a magnificent ball of fire leaping skyward. In slow motion thousands of scraps of wood and metal rocketed into the morning air. Then just as these projectiles began to drop into the sea like hailstones, the sound arrived, a jarring concussive thump that seemed to rock the life raft and flatten Sverre's ears against his head, as thousands of gallons of diesel combusted and blew the *Foremost* to smithereens.

The men watched the fireworks. A grisly black cloud swirled above the disaster and was dispersed by the wind.

"Just like in the movies," said Sverre with a bitter grin. But beneath his smile, he was worried. Nothing brings home how lonely it is in the middle of the ocean better than watching your ship explode. The men took a quick inventory: a few water jugs, some canned K rations, and a flare gun. Not much, especially when you consider that no one was coming to the rescue.

And then, *whooooshhh*.

The awful sound of air escaping from the rubber raft.

They had a puncture.

In 1971, Tom Economou, a Greek immigrant who owned a parking garage in downtown Seattle, was approached with a dubious business proposition. A friend who'd just gone bankrupt looked him up one day and was thrilled to pass on some news.

"The Ballard Smoke Shop is for sale!"

"What's that?" asked Tom.

His friends explained that it was a cocktail lounge up on Ballard Avenue, a rough part of town overrun by fishermen.

"What do you want me to do about it?"

His friend wanted them to be partners. Tom would supply the capital, and the friend would run the place. Tom was willing to take a risk. Sight unseen, he agreed to the venture. But first one more question.

"Where's Ballard?"

The next Friday night, Economou ventured north across the Ballard Bridge to inspect his new property. In his experience, a cocktail lounge was a nice, quiet place with piano music and fancy drinks. That's what he figured he'd gotten into.

The reason I'm telling you about Tom Economou is that the Smoke Shop ended up being the bear's lair for crab fishermen in Seattle. The whole era of big-money crab fishing and rowdy drinking by a rough crew of squareheads—it all went down in the Smoke Shop. In the Viking sagas, you had the Great Halls with goblets of mead wine and whole roasted pigs; in the Hansen saga you have a double Crown Royal straight up at the Smoke Shop.

Tom Economou is not a big guy. He's compact, about 5'8", with a quick smile and an easy laugh. He looks a bit like Al Pacino. His father was a fisherman who used to travel from Greece to Astoria, Oregon, to work on the salmon boats. After World War II, Tom followed his dad to the States. Life on a fishing boat didn't agree with him, which was how he ended up, in his thirties, running a parking garage.

Nothing had prepared him for the Smoke Shop. Its exterior was nondescript, a faded awning that read "Fine Food" and "Amber Room." The restaurant portion was harmless enough—it occupied part of the ground floor of the Princess Hotel, a weekly or monthly rental that was well past its turn-of-the-century glory. Upstairs were a dozen or so single rooms that shared a common bathroom at the end of the hall, home mostly to elderly fishermen and millworkers.

The Amber Room made Economou shudder. It was dark, had no windows, and you couldn't tell if it was day or night once you sank into its depths. It reeked of diesel fuel and fish, and among the howls and bellows of the customers Economou heard not a word of English, only Norwegian. "It was a zoo," he remembers.

Economou made a beeline to a corner table and sat down to survey the chaos. Big burly blond fishermen yelled, fought, bragged, cursed, and drank. "How am I going to get out of here?" he wondered.

By two in the morning on Tom's first night, when the place shut and the bartenders heaved the last of the bunch out the door, Tom sat down with the previous owners to sign the papers. It was empty now, except for a big-headed old man with a crooked neck, who lolled alone at a booth. Fine. Tom and his partner signed the contract and shook hands with the sellers and had a congratulatory drink.

"One more thing," Tom said. "Where are the keys?"

"Keys?" they said. "You don't need no keys." They motioned to the man in the corner. "He keeps the keys."

Tom did a double take. "Who's that guy?"

"That's Marvin. He takes care of the place."

"What am I supposed to do with him?"

"He's included in the deal."

Marvin was disabled and lived in the basement. Each night between 2:00 and 6:00 A.M., Marvin swept the floors and put the chairs and stools in their places. He also counted the tills.

Tom went home to sleep. He was awakened by the phone. He jumped out of bed and answered. It was Marvin.

"One of the waitress's till is four dollars short."

"That's okay, Marvin," Tom said. "Go to sleep."

The next night, the same thing. Marvin was calling to report that the cook had not scrubbed the grill to a perfect sheen.

The third day, Tom pulled Marvin aside.

"You called last night, right?"

"Yes, Mr. Tom."

"Which finger did you use to dial?"

Marvin held up his right pointer.

"You call me again in the middle of the night, I cut that finger off."

Nobody knew much about Marvin. His last name was Sjoberg. He was also known as the Mayor of Ballard. Although he was likely Swedish, Marvin would lead the annual Norwegian May 17 parade—in a tuxedo. In general, he wasn't a fine dresser, yet he refused to buy his shirts at JCPenney's on Ballard Avenue, dismissing them as "junk." He paid twice as much at the Brick Shirt House on Market Street.

Marvin also did a lot of cleaning up down the street at the little triangular park on Ballard Avenue. King Carl Gustav XVI of Sweden had visited the park to celebrate the creation of the Ballard Avenue Historical District. The old Ballard City Hall bell, from back before annexation, hung in a tower. The park came to be known as Marvin's Garden, and Tom wanted to make sure it stayed that way. He ordered a plaque built at his own expense, and hung it in the park. Years later he noticed that it had been removed. He made a stink with the city, and they replaced it with an official sign—MARVIN'S GARDEN—to memorialize the now-deceased Mayor of Ballard.

Tom Economou took over the Smoke Shop, but his partner—the cook—soon backed out. "I didn't even know how to make coffee," Tom says. So he asked his older brother to step in and run the restaurant. Pete Economou was supporting a wife and three daughters by working full-time as a machinist for Boeing making airplane parts. He didn't know how to cook, either, but he wanted to help his brother, and he liked the idea of being his own boss. He began moonlighting in the kitchen, and taught himself to flip eggs and fry burgers. Within a few months, he quit his Boeing job.

The two brothers were full partners, and became a Ballard institution for the next three decades. With Pete manning the kitchen and Tommy behind the bar from 6:00 A.M. to 6:00 P.M., the Smoke Shop was open 365 days a year. "They support me all year long," Pete used to say of his customers. "How can I abandon them on Christmas?"

With Malmen's and Inky's gone, the Smoke Shop became Ballard's Karmoy hub. They called his bar the *Smokesjappe.* "I know all about that stupid little island," Tom says. When a relative from the Old Country was trying to reach a family member in Seattle, they would call the bar. If Tom couldn't understand their poor English, he passed the phone to a regular.

"Oh, you're looking for Hans? He's up in Alaska until next week. You can call back then."

Many of the young fishermen were transient. They spent so much time at sea that they didn't bother to rent an apartment or open a bank account. Tom allowed them to have their mail delivered to the Smoke Shop, where he held it for them. He cashed their checks, and if they were trying to save money, he locked it in the safe like an unofficial bank.

The Karmoyverians were clannish. They stuck together, hired one another, and didn't always mingle with the previous generation of more Americanized Ballard Norwegians. The old guard mockingly called them *Karmoyboos,* and considered them ultraliberals suckled on European socialism. That didn't really make sense, though, considering that the Karmoyverians had grown up in poverty far more brutal than their Americanized counterparts. The newcomers were jealous and competitive, and they didn't always trust one another. Arguments over who caught the most crab would escalate into fistfights. Over time, though, the community of Ballard Karmoyverians built a bond that has endured.

It's time for another update on Edgar. When he started working full time on the *Northwestern,* it changed the dynamic. Since I'm the captain, my word goes, but since he's my smart-ass younger brother, he felt comfortable challenging me. We had plenty of clashes. Once he was trying to wake me up and couldn't do it. The rest of the crew was down in the galley waiting to work, and getting pissed off.

"Is he up yet?" the engineer asked.

"No," said Edgar, loud enough for me to hear. "The lazy fucker won't get out of bed."

That did it. In my family, "lazy" is the worst insult. That pushed me over the edge. I came down, stomping out of my stateroom.

"All I could hear was *boom, boom, boom*," is the way Edgar remembered it. "He came running down the stairs with the look of a man who just woke up out of a coma. He took a roundhouse swing. He gave me plenty of time—I ducked."

Then just to piss me off more, he said, "You want to try that again when you're awake?" Smart-ass.

Another time we were fishing far up in the northwest, two days out of Dutch, in what we call the shitpile. Usually when you flip an opilio, they are white, real clean, and shiny. In this area, however, they were brownish yellow with black spots, barnacle-y, and real ugly. We set up out there and pulled some pots full of eight hundred opies. We boiled a few up and tasted them. They tasted fine, but food has to be aesthetically pleasing, and crab from the shitpile didn't meet that standard. So we were throwing the ugliest ones back. Every pot was full. The table was overloaded, so there were three deckhands sorting crab all the time. I didn't want to slow down to sort this crap. I told them to get the next pot in the block, and if we had to haul it up slow that would be fine—so we'd always have one in the rack, one on the block, and one on the bow. It was relentless.

It was getting toward summer, probably June, and the weather was nice. The sea was flat and calm. Mark, Brad, and Chris were working

Edgar. *(Courtesy of EVOL)*

the pile, and Edgar was multitasking—operating the block while cutting bait with a twelve-inch knife. It looked like a machete. Suddenly
a knot came up in the coiler. With the knife in one hand, he grabbed
the line with the other, bent it back in, and flicked it back in the coiler.
When he slammed down his arm he impaled it on the knife. Then we
heard, "Ah fuck!" Edgar was gripping his wrist with a stain spreading
on his sleeve.

They took him inside and I came down to look. There was deep red sludge oozing from the wound. The muscles and tendons were severed. "Oh my god," I said with a smirk. I never much liked the sight of blood. "What is that shit coming out of his arm?" Edgar turned to me, white as a ghost. He still hadn't looked at it. We held the wound over his head and blocked his eyes so he couldn't see it. Luckily, Mark Peterson the future paramedic was there—he knew how to tend injuries. Edgar healed up OK.

We were two days out of Dutch, but we didn't want to leave the grounds. I didn't want to stop fishing. There was another boat nearby heading to town, so we put Edgar on a skiff, dropped him with them, and they took him in. We kept fishing. Now there were just three deckhands fishing the shitpile. It was crazy. The good thing was that they all got more money, because they took Edgar's share and split it up.

My brother likes to give me a hard time for the television cameras, so in the spirit of brotherly love, this saga wouldn't be complete without me giving one really embarrassing story about Edgar.

One time, we were out by Adak fishing for cod. We had ventured into cod fishing as a way to keep busy and supplement our income in between crab seasons. The crab seasons had become so short, and the prices had dropped, so this new fishery was becoming more lucrative. We basically used the same setup as we did with crabs: dropped baited pots and lured in the cod. We were new to it, didn't know Adam from Eve, and didn't know how to do it efficiently. We had been up for a few days, and I was grinding the crew's titties to the deck. I was very tired myself, nodding off at the wheel. We were fishing just a mile or so from the island, which rises out of the sea like a mountain.

"I need a nap," I told Edgar. "Just forty-five minutes. You want to watch the boat?"

Edgar said he'd do it. He looked fine.

"Are you tired?" I asked.

"No, I'm all right."

Of course, that's what he would say. You never admit you're tired on a crab boat, and besides, he looked fresh as a daisy. So I went to sleep. Edgar peeled off his gear and came in from the cold to the wheelhouse where it was nice and warm. So comfortable. So drowsy. Everybody else was asleep. Later, Edgar tells me, "Five minutes into it, I'm out—dead coma. I don't know how long we jogged for. I woke up and looked out the windows, and all I see is this skinny strip of green on the water. What the hell is that? Kind of odd. And then I see sand. Oh shit. That green stuff—it's grass."

What'd happened was Edgar had ridden the boat up onto the beach. He jammed it in full reverse, which he knew would probably wake me up. Edgar's version is, "I'm praying: *Please don't wake up, please don't wake up.* I've got my eyes closed. We're doing about five knots. And when I hit reverse, the ass-end came around, she turned sideways to the beach, and clunk, we were beached. When something this big and heavy hits bottom, it's a scary, eerie feeling. These boats just don't belong on land."

Of course, he woke me up—as he did everyone. We ran up to the wheelhouse. Edgar was calm about it. He figured that we hadn't hurt anything or anyone—yet. We were just stuck. Edgar said he was more freaked out about getting in trouble than he was having the boat on the beach.

I wasn't so calm. I took one look at our situation and hit the panic button. I started yelling and screaming. He'd beached the damn boat!

Finally we collected ourselves. I tried to maneuver the boat backward, but the wind and waves kept us pinned at a 90-degree angle to the beach. We got on the radio to find help. We were two days from Dutch Harbor, and we wouldn't have been surprised if the nearest boat was ten hours away. But by sheer good fortune, a big cod boat, the *Aleutian Lady*, was just three hours away. The skipper said he was on his way.

In the meantime we tried to pump our tanks to lighten the boat and see if we couldn't float off, but then we just started sucking sand into the pumps. The tide was going out. We were doomed. If we got left high and dry, we could lose the boat. Once a boat this big sinks down in the sand, it's part of the landscape. Permanent. We couldn't back out of it because with the boat lying sideways, the prop wasn't submerged. What's worse, the boat was lying on her side at a 45-degree angle. With each wave, the massive hull heaved up the beach, high and dry, then violently rolled down, like a two-hundred-ton steel rocking chair. We got the crew in survival suits. We readied the skiff with clothes, food, and water. We were all standing on deck, freezing, ready to abandon ship and go to shore.

We were actually damn lucky. Rock reefs loomed on both sides of the beach. We had fetched up on the only sandy spot on the shore. If the *Northwestern* had landed a hundred feet to either side, it would have been dashed to pieces. Edgar told me, "At that point in time, I found God, I guess you could say. I knew that somebody was watching out for this boat and its crew. Definitely."

The *Aleutian Lady* arrived. They fired a gun with a projectile attached to 450 feet of twine that was attached to a four-inch line, which we hauled aboard. The tide was dropping quickly. The captain had no choice but to give it everything. The line tautened as the one boat heaved on the other.

"Back it down," I called on the radio, worried he'd snap the line. "Take it easy."

"Yep, yep," he said. "You betcha, Cap. I'm slowing her down."

But I could tell he didn't. He was just saying that to calm me down. In reality he was cranking on it with all he had. Still we didn't move.

And then, like another miracle, a big swell rose from the bay, like the hand of God lifting us up. The boat rocked unsteadily, then the stern floated free, and was pulled back by the *Aleutian Lady*. The *Northwestern* righted herself, and the prop found water. I jammed it into reverse, and we were free. I've never felt such relief as when I felt that boat floating beneath me. If not for that wave, the family vessel—our family legacy—would still be on that beach today.

Now Edgar was nervous about having to break the news to Dad. Of course I was looking forward to it. By the time we reached a phone at Akutan, Dad already knew all about it. Word travels fast in the Norwegian mafia.

"*Ja*, so, heard your ran it aground!" said Dad. "What the hell were you doing?"

"Sleeping," said Edgar.

"How long were you up?"

"A couple of days."

"Vell, it happens."

Edgar got off easy. Typical for a younger brother.

Overall, it works out well between Edgar and me. I have my job, he has his. There's not a lot of competition between us to be captain. "I'm not really interested in running the boat," says Edgar. "I like being outside. I like being the go-to guy. I just wouldn't be able to stand sitting upstairs and watching what was going on down on deck without being a part of it."

Speaking of stories about younger brothers' mishaps, the last I told you about Uncle Karl was in 1966 when he got married in my parents' new house. Well, his honeymoon didn't go so well. To begin with, he didn't take his bride. Just days after the wedding, he flew to Bristol

Bay for a salmon opener. Karl and his partner, Ray Alfsvag, were fishing a little gillnet boat, a homemade plywood piece of junk they'd leased from a Slovenian. As they hauled the net, the boat started taking on water. The pumps were clogged. Lucky for Karl, some other friends were out. Karl started picking the fish off his boat and tossing them on theirs. The other boats started towing Karl toward shore. The wind was increasing. The boat started sinking, and Karl jumped aboard with the rescuers. Ray was still on the sinking boat, so they told him to let go of the boat and hang on to the line, and they'd pull him in. Instead, Ray let go of the line. He was standing on the boat as it sank. Karl and the others doubled back and picked him out of the water. They returned to fish camp.

The next day Karl and Ray rented a small airplane. Flying above, they saw the boat washed up on a beach on the north end of Naknek. The pilot set them down on the rocky beach and left them. The tide was out. Karl and Ray started cleaning up the boat. They worked all day. The homemade craft was an open boat with no cabin—it wasn't meant for overnight sleeping. They stopped the leak, but they couldn't start the engine. Then the tide came in. The boat started to float. They figured the tide wouldn't be high enough to take them anywhere.

As night fell, the water beneath them deepened—too deep to wade to shore. Still they couldn't start the engine. It was colder than hell. If they wanted to get to shore they would have to swim. The men were soaking wet. They toughed it out all night, just shivering beneath the stars. The next morning the tide went out, and they trudged across the wet rocky pools to the beach and started walking. It was just luck they didn't get caught by a bear. When they came to a fish camp, they called in on the VHF and a plane came down for them.

The next day they returned in a boat. They attached a line to their boat, towed it back to the harbor, and got it up on the dock. By the time they got it fixed the season was over. Karl came home from his honeymoon with no money, no clothes, no nothing, but that didn't stop him from returning to Bristol Bay and becoming one of its top producers. You could even say he had a second honeymoon; years later he bought the boat *Elka*—short for Else and Karl—and he and his wife spent a few summers fishing together. Their time together in Alaska is a testament not just to their skills at sea, but to the strength of their marriage.

As for my father, he was never interested in salmon fishing. He liked crab. In 1967, when his skipper John Johannessen bought his own boat, Sverre became captain of the *Foremost*. It was a great achievement for a young immigrant who'd arrived in America with nothing but his oilskins.

His promotion came at a tough time. After nothing but growth in its twenty-year history, the catch had plummeted from its 1966 peak of 159 million pounds to less than a third of that in 1970. Concerned that king crab were overfished, in 1969 Alaska Fish and Game ordered its first-ever closures, shutting down some fisheries from March through July. Yet there was a silver lining, As the supply tightened, the price shot up. So the fleet's earnings remained stable. Savvy, persistent fishermen could still make a good living, but they had to work harder to find the crab.

For those who survived the slump, a resurgence awaited. Nineteen seventy-one showed the first increase in five years. Having survived the sinking of the *Foremost*, Sverre decided it was time to stop working on another man's boat. Along with partner Dan Pierson and a big loan from the bank, he commissioned a steel crab boat from a boatyard in San Diego. Barely a decade beyond the crushing poverty of postwar Norway, Sverre now owned a boat. He had achieved his dream.

The steel *Foremost* in 1973. *(Courtesy of the Hansen Family)*

In honor of the boat that exploded off Akutan, his new vessel, a steel-hulled crabber, would also be called the *Foremost*. Later, my grandfather told me that it was bad luck to name a ship after one that had sunk. I never learned why my father did it, but as it turned out, there was bad luck in store for the new boat.

With Snefryd three months pregnant with Edgar, my parents flew down to San Diego for the christening. Meanwhile the toddlers, Norman and me, stayed home in Seattle with Grandfather Sigurd, who'd come over from Karmoy to babysit. The Hansens and the Piersons celebrated at a fancy restaurant on the waterfront, and to christen the boat, the wives smashed bottles of champagne across the bow. Mrs. Pierson flew home, but my mom got aboard for the cruise north to Seattle. Between the morning sickness and the seasickness, she knew right away it was a bad idea. She felt terrible. After thirty miles at sea,

Sverre turned the *Foremost* around and dropped her in San Diego, where she caught a plane home.

My dad brought his father on the maiden voyage to Alaska in the new *Foremost*. Don Pierson and John Jakobsen—my first salmon skipper—formed the rest of the crew. They were plagued by bad weather, mediocre fishing, and mechanical problems. "That boat was a misfit," Uncle Karl says. "A hunk of junk," concurs Howard Carlough. "Kind of pretty, but pretty doesn't work." The gulf crossing was un-eventful, but when they refueled in Sandpoint, a port off the Alaska Peninsula, they didn't trim her right. They filled the port tank entirely before filling the starboard, and the *Foremost* just lay down on her side, right there in the harbor. They got it straightened out quickly, but it didn't bode well for her stability.

Grandfather Sigurd was with them for the entire nine months. In order to work legally, he had to immigrate, so Sverre put up the five-hundred dollar deposit. Sigurd was a hell of a fisherman, but he was almost fifty-seven years old at the time, and the work was a bit too demanding. Sometimes he would tend to the galley, heating up left-overs, but he wasn't much of a cook.

"Good stuff, Don," Sigurd said in broken English as he served up some mystery platter to the younger man.

Don took a bite and grimaced. "*No* good."

Sigurd was homesick. He sang the old Bobby Bare song in Norwe-gian, with its mournful chorus: "I wanna go home."

As a skipper, Sverre was like a caveman—barbaric and hard-core. He was military style. When he woke up the crew, they had three minutes to get on deck. If they didn't have their gear on, they heard about it. The crew used to say that Sverre had two speeds—fast and slow—and slow was worn out. One year when Norm and I were little he fished straight through Christmas. His greatest talent as a skipper was not navigation or finding crab—it was running his crew. Run, run, run. You'd work thirty hours straight.

He loved his wheelhouse, and he never left it, even when he was in port. Once he was tied up in Dutch Harbor, watching a guy stagger from boat to boat, and out of the corner of his eye he saw this guy slip and disappear between the boats. He decided to take a look. Sure enough the guy was bobbing in the water. Sverre got the crew to fish him out. Another time during a strike in St. Matthew, when the boats were anchored in the harbor, he was sitting up there in the wheelhouse when saw a head bobbing between the boats, drifting to sea. The guy was drowning. Dad got on the radio and saved that guy's life, too.

As owner of the steel *Foremost*, Sverre earned money even when he wasn't captain. As a result, he no longer had to work the entire year onboard. Instead, he hired Oddvar Medhaug as skipper, and they took turns in the wheelhouse. Sometimes Tormod Kristensen did a spell as relief skipper. The Coast Guard required that skippers be American citizens, but Sverre and the other owners knew they would make more money if they hired a squarehead. It was almost like an artist: If you knew he was good, you could trust the canvas would come out good. So they fudged the paperwork, usually by having at least one American aboard who held the proper license, but the Norwegian actually ran the boat.

The new *Foremost*, as I said, was none too stable. An identical sister ship, the *Aleutian Star*, owned by Sverre's old crewmate Bill Osborne, rolled over and sank in Puget Sound right outside the Ballard Locks in flat calm water. The next year Sverre was steaming across the Gulf of Alaska, hundreds of miles from land or another ship, with all the empty pots stacked on deck. A storm descended and the boat was knocked on its side by a monster wave. The men held tight, unsure whether the next wave would right the craft or capsize it, but the *Foremost* held its position on its side. Finally the men ventured out on the near-vertical deck with knives, and cut loose the pots, sacrificing the pots as well as the shots and buoys stored inside, in order to save their lives. This desperate act is called "suitcasing." It worked. The pots slid off the deck with a grinding thunder and splashed into the sea. The boat sprang up out of its list. Miraculously, everyone was safe. Sverre had lost a hundred thousand dollars' worth of gear, but he'd survived. They motored up to Dutch Harbor, where he bought new pots for the season.

Spooked by his brushes with death, Sverre outfitted the new *Foremost* with the latest in safety equipment: quick self-inflating life rafts and rubber survival suits that kept a man afloat and alive for up to four hours in the freezing sea. It may well have been the first boat in the fleet to employ the expensive—and, to some eyes, ridiculous—contraptions.

The mishaps continued. Another winter Sverre was fishing off Adak beside his friend Jan Jastad. Jastad was low on diesel, but the only fuel for sale on the navy base was a product called JP5, which fishermen considered too dry for their engines. So Sverre agreed to pump a few thousand gallons of his own diesel into his friend's tank. The engineer, a guy named Black Jonas, pumped the fuel out of the *Foremost*. Instead of draining the upper tank, he pulled from the lower.

Sverre and the *Foremost* headed back to Dutch Harbor, coming around Cape Cheerful. With its belly tank drained, the boat lacked ballast, and was top-heavy. The seas came up and tossed the *Foremost* around. It felt a bit loose. Sverre checked the engine room. Everything was hunky-dory. Then, just as he climbed up into the wheelhouse, the boat rocked mightily. He thought she going upside down. Instead she listed so far that the wheelhouse window submerged in the sea. The boat popped back up on its own. When Jastad saw Sverre back in Dutch Harbor, his friend looked haggard.

"I've never been so close to death," Sverre confided. It's the only time Jastad ever saw his friend truly rattled.

During the months that he was home, Dad was a family man. He came to our soccer games, and took the family on road trips or up to mountain lakes to fish for trout. He gave us boys haircuts using an electric clipper he'd been given for Christmas. He thought he was Jack Johnson down at the corner barbershop, and we kids were his guinea pigs. He lined us up on the patio with towels around our neck, and we were terrified. "Three boys, three shaved heads," he said. The next time, I wasn't up for being his test piece, so I ran off. When I finally returned after dark, he got a chuckle out of it. Once he wanted to go watch a soccer game in Oregon, so he drove me down to Sea-Tac Airport, bought two tickets, and we flew down to see the game.

He also had a side that was a sea captain through and through. When he came home, he'd start barking orders to us kids. One year he rented a satellite dish to watch the World Cup. They delivered it on a trailer that parked in the driveway. He couldn't be bothered by anything when he was watching soccer. Norman crinkled up a Coke can and Dad told him to knock it off. Norman decided to push it, and crinkled the can again with a smirk. "That got a boot in the ass real quick," Norman says now.

When he was away, we kids would get in trouble, and Mom would say, "Just wait till your father hears about this." He might not be home for five months, so we didn't worry. When his arrival got close, we'd be sweating bullets, because we'd got an F on a report card or burned down a tree in somebody's backyard. One year when he was away my mom bought a brand-new couch from Sears. It was orange velour. We didn't live high on the hog. My parents came from very humble beginnings, so when they made a dime they knew they'd damn well better save it. She loved that couch. One Saturday morning, the day she slept in, Norm and I woke up early and smeared the thing with Vaseline. I don't know why. Just because. Mom got up, took one look at the couch, and sat down in the middle of the floor and wept. Our family had worked so hard for what we had, and now we kids were ruining it. A kid who grew up in Depression-era Karmoy surely wouldn't have done something like that. Eventually she had some pros come out and clean the thing. She never told Dad about it. "If she had laid down the law a bit earlier," says Edgar, "I might not have turned out the way I did."

Meanwhile, as Dad and his brother aged, they proved the old saying that blood is thicker than water. They didn't hang out every day. They hadn't worked together on a boat since 1966 when Sverre first took over the *Foremost* and hired Karl on deck. They were intensely close and competitive, as only brothers can be. They showed their affection by teasing and bickering. Now and then they'd get under the other's skin in a way you can only do when you really care about someone.

The stories of their rivalries and kidding around are still repeated. In 1974, Karl had finally returned to crab fishing, becoming captain and part owner of the *Ocean Spray*. One stellar season Karl was the top skipper in the fleet. His partners presented him with a watch engraved with a nickname that Karl would carry his whole life: The Champ. One summer Karl brought the *Ocean Spray* into Dutch Harbor from Kodiak. On they way, he baited his crab pots and set them. He was still somewhat new to crabbing, and not as confident as he was on a drag boat. When he got to the Elbow Room, he ran into Sverre and another highliner, Buddy Bernstein, and told them where he'd set his gear, hoping he'd chosen a good spot.

"Hey, Champ," Sverre teased, "I'm gonna buy your boat for spare parts."

Sverre informed him that a strike was on. According to the rules of the strike, all the crab fishermen had removed their bait boxes, so that when the strike ended, they would all start on equal footing.

"You didn't bait it up, did you?" Sverre said, egging him on.

"No," Karl lied.

Sverre and Buddy burst out laughing.

"How the hell can you expect to catch any crab with no bait?"

"I thought you said we weren't allowed to bait!"

Sverre and Buddy busted a gut, teasing Karl as he tried to figure out what he was really supposed to do. This time, though, Karl got the last laugh. By the time Sverre finally went to re-bait his pots, Karl's had had a long soak, and were already brimming with crab. He filled his tanks long before his brother.

Early one morning in April of 1977, around the time I was home celebrating my eleventh birthday, the steel *Foremost* was bucking fierce winds in the Bering Sea, halfway between the mainland and St. George Island in the Pribilofs. Oddvar was the skipper, along with four crewmen—two old hands from Karmoy and two greenhorns from New York he'd picked up on the docks at Dutch Harbor. Oddvar was asleep, with one of the crew on watch. He awoke violently when he got thrown from his bunk. The *Foremost* had keeled over 90 degrees—and stayed there.

Oddvar rushed to the wheelhouse and radioed a Mayday. The panicked crew ventured out on the bridge in the 50-knot winds. The life raft was already underwater. There was no way to deploy it. The men huddled in the sideways pilot house. Oddvar broke out the survival suits, the newfangled gizmos that no one had ever worn, or had much faith in. Water was pouring into the galley and the ship began to sink.

Suddenly Oddvar made a grim realization: there were only four suits. Since he'd taken a second greenhorn on this trip, there were five men. As skipper, he knew he was the odd man out. The captain goes down with the boat. Oddvar remembers, "I was the last one. I thought I was gone."

Then one of the greenhorns spoke up: "I got my own!"

Sure enough, the kid had purchased a survival suit in Dutch Harbor and packed it in his seabags. They wasted no time, stuffing their legs into the thick black neoprene, then arms, then stretching the neck gaskets and hoods over their heads. The low side of the pilot house was flooding now, so they made their way to the bridge. The men clung helplessly to the rail as the boat flooded with water. The hatches to the engine room and lazarette succumbed to the building pressure, and seawater rushed into every cranny. One by one the men lowered themselves into the icy water, gripping the rail for final comfort, floating beside the doomed vessel. A survival suit is more like a wet suit than a dry suit—it heats the water that seeps in, but does not prevent it from entering. It keeps you alive—not warm. The men shivered in the suits as the icy water seeped along their spines, down their thighs and calves.

Then the *Foremost* was gone. With a sickening gurgle and whoosh, hundreds of tons of steel dropped to the ocean floor. The men held on to one another in a circle, bobbing in the Bering Sea. There was nothing to do now but wait. The manufacturer of the survival suits had claimed that you could live four hours in one of the things. A Coast Guard rescue would take much longer than that.

Ten minutes. Twenty minutes. Thirty minutes.

Suddenly a boat approached. It was another crab vessel, the *Sea Venture*, that had heard the Mayday. Its skipper was Oddvar's old friend Chris Knutsen. He'd steamed to the spot and seen nothing but a floating buoy. He followed the tide and finally reached the five desperate specks bobbing in the infinite sea. They were pulled to safety. "I wouldn't be here without that survival suit," says Oddvar.

When Dad got the news that morning, he was oddly relieved. Though his boat was lost, his close friend Oddvar had survived, along with the rest of the crew. He called Oddvar's wife to deliver the news.

"Everyone is saved," he began.

It scared the hell out of me. That morning when I heard the news, I went to school as usual, but before the bell rang, I had what qualifies as a total breakdown in the cafeteria. We had lost our boat, our family livelihood—and I didn't know how we'd go on. It was like our house had burned down. I stood there by myself, sobbing. A teacher asked what was wrong, and when I told her, she sent me home for the day. The cause of the sinking of the *Foremost* was never established, but ultimately, we got off pretty easy.

"It was a misfit," Oddvar says of the steel *Foremost*. As it turned out, the whole line of these boats was flawed. Fourteen built, fourteen sank. "It's a good thing it went down without people on it. Thank God for that."

My dad was not cowed by the disaster. Days later he walked into the Marco shipyard on Salmon Bay and cut a check for the down payment on a new boat. For the rest of that summer, Sverre was out of business, anxious for the arrival of his new craft. He waited and waited. One day while driving around Dutch Harbor with Charlie McGlashan, he'd taken a look at a old shipwreck in Captain's Bay.

"What's the name of that boat?" he asked.

"The *Northwestern*," said Charlie.

Sverre produced a notepad from his coat and scribbled the name on a list he'd been accumulating. "That's a great name for a boat."

For his new vessel Sverre would not scrimp. He would not build another misfit. This time he had commissioned the best boat on the market.

In 1966 a ninety-six-foot steel boat, the *Peggy Jo,* was launched by the Martinolich Shipbuilding Corporation of Tacoma. With its raised forecastle to resist icing, a forward house, and a hydraulic crane and power block on a long wide deck, it was the first boat designed expressly for crab fishing with rectangular steel pots. The next year, Pacific Fishermen in Ballard rolled out the *SeaErn,* a similar hull but with the house aft and the deck forward. The buyers were Norwegian-Americans. The *Peggy Jo* and the *SeaErn* launched a golden era of American shipbuilding. Over the next fifteen years, two hundred crabbing vessels were built in the United States, most of them in Puget Sound.

"By then we wanted to go to Dutch Harbor, and Adak, and the Bering Sea, and it wasn't going to be done on a fifty-four-foot boat built in 1917, I'll tell you that," said Lloyd Cannon.

This might sound like a bunch of technical jib-jab, but it's important. Without these innovations, the crab industry would have remained a tiny business. Just as the Viking naval architects allowed the longships to roam the northern seas, and just as breakthroughs in European vessels allowed the explorations of Columbus and Magellan, so did the brilliant designs of a few shipbuilders enable the golden era of crabbing. My family saga is tied up with theirs.

The shipbuilder that best rose to the challenges was just across Salmon Bay from Ballard: Marine Construction and Design Company—Marco. "Around nineteen sixty-nine we got the first order for a crabber, the *Olympic,*" recalls Peter Schmidt, the company's founder. "We designed a totally new boat. A ninety-four-footer. The boat was very successful. We immediately got more orders. We'd sculpted it out right. You have to make it super stable, because of the load they put on in crab pots, and because of the icing conditions."

Marco boat #189, the *Olympic*, launched through the Ballard locks in 1969 and became the prototype for a generation. Designed by naval architect Bruce Whittemore, the company's first-ever crabber was ninety-four feet long with a twenty-five-foot beam, with the house forward and the working deck to the aft. A finely sculpted and all-welded steel bow rose high out of the water, affording a clear unobstructed view from the raised wheelhouse. The steel plates of the hull were lapped over one another, just as the Vikings had done with wood planks. The ship had two tanked holds, that could carry 110,000 pounds of crab and another 7,400 cubic feet of dry storage. Fresh seawater constantly circulated through the tanks by a pump and refrigeration system. Outboard buoyancy tanks ran the length of the holds for added stability. To reduce icing, the typical mast with cable stays was replaced by a free-standing tripod mast. As Marco began building more ships in the *Olympic* model, standard equipment included backup electronics in case something malfunctioned at sea: two radars, two depth sounders, and three radiophones.

Norwegians Harold Hansen, Sam Hjelle, and John Sjong bought the *Olympic*. Six sister ships followed in quick order, and were snapped up by owners whose last names—Peterson, Kaldestad, Hvatum, Hendricks, Gudjonsson—reveal that to a man they were Norse. Of the thirty crabbers completed in the late sixties, eleven were owned by squareheads.

The Marco hull length was gradually extended to 108 feet—with a crab capacity of 170,000 pounds, and by 1974 another fourteen vessels had been delivered. Kaare Ness began building his rich fleet with Marco boats: the *Royal Viking*, the *Pacific Viking*, the *Royal Atlantic*, and the *Nordic*

Star. Sverre's friend John Johannessen bought a Marco crabber, as did his old boss, Howard Carlough, who commissioned the *American Beauty.*

The Marco boats have a stellar record (knock on wood). "Fortunately, we built a total of forty-nine crabbers between ninety-four and one hundred twenty-three feet, plus the *American No. 1*, which is one hundred sixty feet, and we haven't had one of our vessels lost to stability problems," says Schmidt. "We feel very lucky about that."

Though several Marco crabbers have sunk due to running aground or captain error, none has capsized. Capsizing causes the most catastrophic loss of life because the crew has no time to deploy a life raft or survival suits, as they would on an upright sinking ship. A crewman on the Marco boat *Westpoint* was swept overboard and drowned. Although he had no family, the man's church, claiming that as his spiritual brothers and sisters they had undergone much pain and suffering, sued the boat's owner for damages. The owner's lawyers set out to prove that by purchasing a Marco, the "Cadillac of the Fleet," the owner had done his due diligence. Other shipyards, notably Bender in Mobile, Alabama, had a poor track record. Their ships, designed for the Gulf of Mexico, capsized in Alaska. Lawyers introduced as evidence a little ditty that had been sung over the years by fishermen in the Elbow Room and the Smoke Shop. Sung to the tune of Janis Joplin's "Mercedes Benz," it went,

> *Oh, Lord, won't you buy me a Marco crab boat*
> *My friends all fish Benders, they don't stay afloat*
> *Fished hard all my lifetime singing this sorry note*
> *Oh, Lord, won't you buy me a Marco crab boat*

The case was dismissed.

By the midseventies, the crab resurgence had become a bonanza.

This new Alaskan gold rush was unprecedented, the unlikely result of colliding historical factors. First, the technological advances pioneered in the sixties, like steel pots, power blocks, and loran, became standard, greatly increasing the tonnage of crab harvested. The addition of sodium lights on deck allowed nonstop fishing in an ocean that was black for eighteen hours a day. Second, with the influx of capital, a new fleet of steel boats built specifically as crab vessels replaced the old wooden trawlers and seiners that had been converted. Third, more processing factories were built in Alaska, increasing the capacity for canning and freezing and quickening the flow to the marketplace. And last, as boat stability increased, crabbers could push farther offshore from the traditional grounds around Kodiak and the Aleutian Islands. One season, as most skippers were hauling a million pounds of king crab, a Ballard captain named Chris Paulsen just couldn't seem to get lucky. He'd caught less than 70,000 pounds. As the rest of the fleet steamed for Seattle with their fat checks, Paulsen couldn't even afford the fuel to get home. He couldn't pay for his salt. He had to stay. So he went far north into the Bering Sea, beyond the reaches of the old wooden fleet. He struck gold and he filled his tanks. After that, he became one of the top highliners, and the offshore expanses of the Bering Sea became the destination of the whole fleet.

Seeing the profits pulled from the Bering Sea, banks were eager to lend money for new boats. Fishermen were also able to take advantage of the Capital Construction Fund, a government incentive to expand the American fleet. When owners showed huge profits, they could avoid paying income tax by investing their money in building or rebuilding a vessel. You either paid the government, or you built a new boat. It was a no-brainer.

Everything was booming: Greenhorns became full-share deckhands, deckhands became skippers, skippers became owners, and owners became diversified businessmen and bought more boats and expanded into processing, canneries, and real estate. Meanwhile, the boats were hauling in millions of dollars' worth of crab per year. Owners paid off their boat mortgages in the first year of operation. In 1975, Kaare Ness paid off a boat mortgage the same day it was approved.

Marco couldn't build their boats fast enough. They laid a new keel every six weeks. Even people who didn't have the money to buy a boat threw their name on the list. In 1974, Uncle Karl was approached on the docks by a friend looking for a third partner on a Marco boat.

"You're crazy," Karl said. He had just bought a new house. "I don't have any money. Hardly enough for groceries."

Karl was still dragging and gillnetting and hadn't tapped into the riches of crab.

"You don't need money," said his friend, who was a partner with Ken Peterson. He already owned two Marco crabbers, the *Ocean Spray* and the brand-new *Ocean Harvester.* "I don't have any, either."

"What kind of boat?" Karl asked. He figured his friend was talking about a small gill-netter.

The friend pointed across the water at the 108-foot *Ocean Harvester.* The first of Marco's 108-footers had sold for half a million bucks, and the prices were jumping.

"I already told you." said Karl. "I don't have any money."

"Neither do I. The Japanese will back us."

Karl was doing as well as anyone as a drag-boat captain. If everyone else was making a fortune on crab, why couldn't he?

"What the hell?" he said. "Why not?"

They figured by the time the boat was built, they would find funding. So Karl and his friends signed a contract. While they were at the Key Bank office in Ballard, working at the details, Sam Hjelle walked in. He had a contract for another Marco vessel.

"I could have sold my contract today for a hundred thousand dollars," said Sam.

Ken Peterson was dumbstruck. People were so desperate for a crab boat that they were paying a hundred grand just for a *spot in line;* they would still have to pay full price when the boat was finished.

Well, this sounded like a pretty good deal to Ken. Later in the year, Karl's friends invited him to become a partner on their older Marco boat, the *Ocean Spray.* Suddenly it didn't seem all that important to have a second boat. So when their Karmoy buddy, Gunnleiv Loklingholm, got eager for a boat, Karl and his partners sold him their contract for a cool hundred grand. The next time Karl arrived home from Alaska, there was a message from the bank to go pick up the check. When the manager delivered a draft for $33,000, Karl could hardly believe it. Free money! He marched right down to the Smoke Shop and bought a round. Tom Economou sat down and had a drink with him.

"Hey, Greek, can you cash a check?"

"Sure thing, Champ."

Karl pulled the check from his coat pocket. Without even looking at it, Tom took it back to the safe.

"You better look at it first!" Karl called after him.

When we talk about the beauty of a boat, or the riches it wins us, we forget the most important thing: Does it stay afloat? Ultimately what made the Marco and Martinolich boats such feats of engineering was that they tended to stay upright, no matter what ring of hell you sailed them into. And you can never really appreciate a boat that stays up until you've been on one that goes down. I've been lucky enough, knock on wood, to never have been on board a boat that sank. For that I have to thank my father and his crews, for all the risks they took when the learning curve was steep and the consequences high.

On November 14, 1977, Marco delivered boat number 342, the 18th steel crabber of the 108-foot class, with the house forward and the entire deck clear for stacking pots. The three holds totaled 7500 cubic feet, enough to carry 85 tons of live crab. The engine room hummed with a turbocharged, after-cooled Caterpillar D398 diesel, linked to a Caterpillar 7251 hydraulic reverse/reduction gear, producing 850 horsepower to drive a Coolidge 80-inch, 3-bladed stainless steel propeller that provided a cruising speed of 12 knots. Two auxiliary Caterpillar D3306 engines powered the 135-kilowatt generators. On deck was an all-hydraulic system that included a Marco pot hauler, dumping rack, anchor winch, and boom winch and an 8-ton Rowe crane. With the $200,000 of crab gear, the total bill for the new boat ran $1.5 million.

The *Northwestern*'s white hull shimmered across the water, the name painted in bold blue. Sverre's initials SH were lettered across the bow like a naval insignia. Usually such christenings took place in summer at fancy dockside clubs—with lots of crab served—but the boat was behind schedule and king crab season was already under way, so Dad held the ceremony at Marco's shipyard. A star-spangled flag hung from the ship's rail. Sverre wore his best blue blazer, slacks, and a tie. Mom wore a stylish black turtleneck beneath a white jacket with a corsage pinned to the lapel. We boys wore our standard issue 1970s puffy striped parkas and blond bowl cuts. Sverre hired Tommy and the barmaids from the Smoke Shop to serve drinks.

The ceremony also had a serious side. Although not a regular churchgoer, Dad was a deeply religious man. He asked his friend Soren Sorenson to bless the new boat. Sorenson led us in the Lord's Prayer. Then he asked the Lord to watch over the boat, keep it afloat and safe, and let it do well to provide for our family.

"We never know if we're going to live or die," Soren said. He spoke of praying for his father and grandfather when he was a boy in Karmoy. "We've all been in tough situations on the water. My grandfather

The family celebrates the christening of the *Northwestern*, 1977. *(Courtesy of the Hansen Family)*

used to say, 'Make sure the Lord is on your boat. He is your pilot.'"
After he was finished, my dad and the other old fishermen hugged
him, with tears in their eyes.

My dad was so proud that day. There were men richer than he
who owned boats outright and leased them to skippers. There were
captains who owned a share of the boats they operated, as Sverre had
done with the steel *Foremost*. As far as I know, by purchasing the *Northwestern* without partners, Dad became the first captain in the Alaska
crab fleet to own his own boat outright. It was a great achievement.

The next week was a frenzy of preparation to get to the Alaska.
The king crab season was already under way. Before they launched,
my mother chose a picture to hang above the table in the galley. It's a

well-known portrait called "Grace" taken in 1918 by Eric Enstrom, a Scandinavian-American photographer. It shows an old, white-bearded, saintly man praying before a loaf of bread. It's a profound image of a man giving true thanks to God. I wonder if it reminded her of the days when my father used to bake bread each night before heading out of Dutch Harbor. At any rate, as launch day approached, the picture had not been hung. It sat on the table, a low priority for the busy crew. My mother protested.

"We'll hang it on the way up," said one of the deckhands.

My mother put her foot down. "This boat is not going anywhere until that picture is hung."

My father mounted the photo in a frame on the wall. Later that day, the *Northwestern* passed through the Ballard Locks and steamed across Puget Sound, on its way north to the Bering Sea.

8

BOOM AND BUST

Air hissed from the punctured raft. The men scrambled over one another, scouring the rubber ring for the hole. At this rate they had only a minute to stop the leak before their raft deflated into a useless rubber sack. Then Krist found the culprit: an air valve between his legs, protruding from the tube, just like a stem on a car tire.

"The valve!" cried Krist. "It's leaking!"

"I can see it's fuckin' leaking!" Sverre snapped. "Put your finger in it, for God's sake."

Krist plugged it up with a finger. They looked at Leif Hagen. It had been his idea to pull this raft out of the galley to where they could deploy it, and they'd thought he was being overly cautious. Now Sverre had to admit that he was grateful. Then came a sound, *Whoosh!* There were *more* leaks. The men looked around. There were identical valves spaced evenly around the raft. They were *all* leaking. Goddamn Leif and his stupid raft. The damn thing was *defective!*

"Plug up the others!" screamed Sverre.

"I've only got two hands!" Krist shot back.

"I don't give a damn how many you got—put them in there!"

Now all four men flopped to their knees in a puddle of freezing water, lunging with their hands to mash their fingers over the valves.

"Hold your fingers there!" Sverre screamed, now furious that he'd made the decision to abandon ship, the rest of his life measured in minutes as this goddamn rinky-dink useless piece of rubber delivered them to a humiliating death. "Plug it up! Don't let go!"

Suddenly the hissing stopped. The men looked at each other in shock. They examined the valves. Then Krist burst out laughing. Then Leif, Magne, and, finally, Sverre joined in. In a moment they all understood.

There was no leak. The valves they were fingering actually served a purpose. They were purge valves, meant to release a bit of air if the rubber tubes were overinflated. When the four men leapt aboard, their combined bodyweight had overstressed the rubber, and the valves had automatically released the excess air. Now that the air had been released, the purging ended. One by one, they cautiously removed their fingers. No whooshing. No leaking. No imminent death. The men howled hysterically, choking for breath. After two hours of panic, here they were, madly plugging their fingers into a nonexistent leak. Sure, it was an awful predicament, but they had to admit it was funny.

Captain Sverre and his men took stock of their vessel. It was the latest in life raft technology, with a rubber tent overhead to keep out the rain and wind. On the floor was an inflatable cushion to keep them from the frigid puddle that accumulated as waves splashed over the rim. They were four men without a piece of raingear between them. Magne Berg wore the only pair of boots. They had no radio, no motor, and no prospects of rescue. No one's cigarettes had survived the ordeal. As their heart rates returned to normal after the morning's firefight and paddling sprint, Sverre could only think of one thing: the cold.

The December day hovered at just about freezing—which in itself wasn't too horrible—but the 40-knot winds dropped the wind chill to well below zero. Without a word, the wet men huddled together on the floor cushion. There was really nothing to talk about. They could holler and scream as much as they wanted, but no one could hear. How long could they survive? Probably a day or so before hypothermia shut down their hearts. Off to the south they saw the icy cliffs of Akun Island. The current was pulling them past the island into Unimak Pass. Each knew from a life on the ocean that their meager paddles were useless against the powerful current. Besides, shore offered them no solace. The cliffs rose directly out of the sea and they would be smashed up on the rocks by breakers if they got close. Even if they managed to land safely, there were no towns on the island to get help. Might as well drift.

"Now what?" said Magne.

"We wait till dark," said Sverre. "Then we fire off a flare."

"To who?"

The question wasn't worth answering.

The *Northwestern* arrived in Alaska in 1977 just in time for the Golden Era of crab fishing. The boom that had begun in the early seventies was now an explosion fueled by a pair of political decisions. In 1974, federal judge George Boldt ended years of lawsuits when he ruled that Native Americans had a right to 50 percent of all fish caught in the coastal waters of Washington State. The Boldt ruling effectively halved the quota of the existing commercial fleet and delivered the final blows to the fishery that was already in decline after nearly a century of profit. Seattle's commercial fleet took the news in stride and headed north to less-regulated Alaska.

Two years later, the U.S. Congress passed a law sponsored by Washington's own Norwegian senator, Warren Magnuson. Since the advent of commercial fishing, American vessels had competed with foreign fleets—mostly Japanese, Russian, and Canadian—for the prize catch off American shores. The Magnuson Fishery Conservation and Management Act of 1976 ended that. It closed all waters within two hundred miles off the Unites States coast to non-American vessels. With Washington's waters basically shut down by the Boldt decision, the Magnuson Act's largest effect was on Alaskan waters, where the competition was now eliminated.

The race was on. The new industrial-strength crabbing vessels converged on Dutch Harbor, with hundreds of pots stacked on deck with hydraulic cranes. The numbers were staggering. Pan Alaska Fisheries CEO Ron Jensen remembers that his Dutch Harbor plant processed more than one million pounds per day. UniSea bought Vita Seafood Products with $19 million of borrowed money in January 1976, and paid off the loan in nine months. The catch climbed steadily, in 1979 reaching 149 million pounds, just short of the record set thirteen years earlier. However, the price of crab in 1966 had been about ten cents a pound; now it was a dollar. So the 1979 earnings were $150 million, a tenfold increase. In 1979, Marco delivered the 160-foot *American No. 1* for the jaw-dropping price of $7 million.

The same year, flush with cash and confidence, Sverre and four Karmoy friends—Gunnleiv Loklingholm, Sigmund Andreasson, Borge Mannes, and Arnold Rasmussen—started a processing company, Aleutian Island Seafoods. They bought from the U.S. Navy a 289-foot refrigerated cargo ship called the *Aludra* and rechristened it the *Aleutian Monarch*. They hoped the boat would help them climb the next rung of industry. They also bought a vintage stripped-down cruise ship, the *Xanadu,* to house the workers. They tied the boats side by side and processed crab—a floating factory. My dad sold our house to Oddvar and moved us to a bigger, nicer place perched on a terrace with a commanding view of Puget Sound. Like the rest of the Ballard skippers, he bought himself a Lincoln Mark V, a huge low-riding luxury car. Each Sunday, the parking lot at Rock of Ages Lutheran church was thick with Lincoln Continentals.

The industry became more competitive. Instead of limiting the number of days in the season, Alaska Fish and Game limited the overall catch. Each captain was required to report their numbers by radio, and when the bureaucrats tabulated that the quota was met, the season was over. The system—known as the Derby—worked fine when there were a hundred or so boats in the fleet. There was plenty of crab to go around, and plenty of time to catch it. Then, as the lucky few got rich, more and more boats steamed north to get their share. There was better gear and faster crews. With sodium lights installed onboard, the crews worked around the clock. The time required to catch the quota tightened from nine months to three months to one month. The difficulty and danger of the job increased, and so did the money. If they could catch in a month what used to take six months, they still earned the same cash.

Here's another way to look at it: If you set ten kindergartners loose on a hundred pounds of candy, they'll binge themselves in their own time, and probably won't cut each other's throats. There is plenty to go around. If you allow a hundred kids, it gets competitive. Now these kids will be throwing elbows, pinching, and biting in a mad scramble to get the most. It's human nature.

After the staggering 1979 king crab season, everyone with a boat that floated wanted a piece of the action. For 1980, Fish and Game announced a quota of 130 million pounds of Bering Sea red king crab—the highest in history. (An additional 30 to 50 million pounds of other types of king crab were caught in different places, but the big bonanza was Bering Sea reds.) That fall, 230 crab boats—the most ever—arrived in Dutch Harbor hungry for loot. At the stroke of midnight on opening day, having already motored to the crab grounds, the boats dropped their pots.

Sverre had come up with a new innovation. He knew that he could catch more crab than his tank would hold. He wanted to bring in more crab per trip, so he brought several dozen giant plastic tote bags so that he could "deck-load," that is, carry an additional ten thousand pounds of crab.

The fishing was fast and furious. After a mere twenty-nine days, the fleet had captured the 130 million pounds of king crab, bringing the year's total of all king crab species to 185 million, smashing the 1966 record. The fleet earned a whopping $175 million. Dutch Harbor dethroned San Diego as the country's Number One fishing port, while Kodiak was Number Three. Deckhands made $50,000 in four weeks, which in today's dollars would be about $130,000, or a tidy $4,500 per day.

"I made more money in one year than my father had made in five years," remembers Bart Eaton.

"A laboring man's answer to prayer" is how deckhand and writer Spike Walker remembers that 1980 season. "One of those freakish moments in history when a working stiff could get ahead for good with nothing more than luck, ambition, and a sweaty brow."

Many of the instantly rich fisherman blew their cash on gold-nugget watches, partying, sports cars, and spontaneous jaunts to Maui. The Elbow Room was pulling in a million dollars per year.

"We set up a jewelry store," remembered Dick Pace, founder of

UniSea. "We would only sell high-end Seiko watches, inlaid with gold nuggets and all the gaudy type jewelry that you can imagine, watches that would wholesale for a thousand dollars. We'd sell them for three thousand. We couldn't keep the shelves full."

The crab fishermen were in a bubble, and none could see its end.

Managing their vessels from home gave the Karmoy boys more free time. Meanwhile, Ballard began to fall into decline as more and more families moved north to the suburbs. The streets were dirty and unkempt. The family-owned businesses began to shutter as shoppers headed to the malls. Even the venerable neighborhood JCPenney foundered, unable to compete with its larger sister store in the Northgate Mall. The fishermen still loved Ballard, though, and frequented the bars and restaurants, especially the Smoke Shop.

"We invested a lot of money there," says Uncle Karl, "but we never saw no dividend."

The Smoke Shop's regulars were wealthy boat owners whose employees and grown sons operated their boats most of the year. The "Norwegian mafia" as they were called made deals, bought and sold boats, hired and fired captains and crew, all in the bar's dark and smoky confines. When Sverre hired Marco to lengthen the *Northwestern* by ten feet, the manager walked over to the Smoke Shop to have Sverre sign the contract.

My father was now a successful businessman, confident and brash. One day while running his boat into the fueling station at Ballard Oil, he knocked it against another vessel, the *Lady Jane*. Dad was always considered a throttle jockey—full speed forward and full speed back when he was tying up. The collision dented the other boat and caused quite a commotion on the dock. People were yelling at him. Sverre climbed off the boat and strolled toward the street as if he hadn't heard a thing.

"Sverre, what are you going to do about this?" someone called after him.

Sverre lifted an arm as defiantly as a matador.

"Send me the bill!" he hollered, and strutted up the street without even glancing over his shoulder.

Sverre could be stubborn and proud. He was the type who would bellow out, "There are only two types of people in the world: Hansens and those who want to be." At times he stuck to the smallest principles. He had a dispute with UniSea cannery over a thousand dollars' worth of bait, and he never did business with them again. Once he hired Tor Tollessen, who'd grown up next door in Karmoy, to install all new electrical on the *Northwestern*. One of the new radios didn't work, which was 2 percent of the whole job. The boat was in Dutch Harbor and Tor gave an order to fix it, but his guy was too busy, and it didn't get done.

Sverre called and said, "That radio isn't fixed."

"Well, it's supposed to be."

"Well, it isn't. So take that frickin' thing and shove it."

So Sverre brought the radio in and Tor refunded the money.

"Of course, I should have never done that," Tor recalls. "He didn't want the money back. He just wanted the radio fixed. It was an insult to him. I took him on his word, but he really just wanted to show that he was upset. Officially, I never was allowed to step foot on that boat again, but every time we had a drink together, we made up. Of course it's a terrible thing to have pride, but it's understandable when you come from the hardship of Norway, in the days after the war, when it was almost a Third World country."

Even at the height of his success, in many ways Sverre remained a simple man from that tiny fishing village. He wore an old farm hat with a shipyard logo, and the more drinks he had, the farther to the side the cap got cocked. When the bait arrived, he ordered his crew to cut out all the codfish tongues, the soft pieces of meat under the mouths. He'd fry them up and feast. One time he was motoring home across the gulf at the end of the season with his friend and mechanic, Mike McCool. Every night Sverre would cook up herring and potatoes. Sverre was a damn good cook. As he was firing up the stove he sent McCool out to the freezer to get the herring. McCool looked around but didn't see anything. It was just frozen boxes of bait.

"There's no herring in there," McCool said. "Or at least I couldn't find it."

"I'll get it myself, you dummy," said Sverre.

He marched out to the freezer and returned with a cardboard box. The label read: BAIT - NOT FOR HUMAN CONSUMPTION. For the rest of the voyage McCool ate only potatoes.

As ships and technology improved, fishermen assumed that the business would become safer. They were wrong. Actually, some of the newly designed boats—like the steel *Foremost*—proved unseaworthy. And naturally, skippers in new steel boats thought they could take more chances than they had in the wooden craft. Some of the new captains, battlefield-promotions with less experience than the old salts, lacked the skills and judgment of their elders. The megacrabbers with six hundred pots proved unwieldy. As experienced skippers stuck to the smaller boats, the captains of these behemoths were young guys clever enough to get the required tonnage licenses. Many of the boats were run aground, or into each other. The number of deaths skyrocketed, earning crab fishing its infamous honor as "America's deadliest job."

"Back in [the old] days, the boats didn't sink too often, because everyone was scared," said Bart Eaton, the highliner who went on to be a seafood executive. "You always figured they were going to sink, but they didn't sink. Now you get the big steel rigs up here, with all the electronics, insulated with all the communications, they're flopping over left and right. I think being scared probably served everyone pretty well."

In 1980, twenty-eight crab crewmen were killed. In 1981, the deaths continued. The captain of the scalloper *St. Patrick* and his nine crew members abandoned ship as it foundered on rocks and flooded with water. All but two perished. In a horrible twist, the *St. Patrick* never sank. All hands would have survived if they had just sat tight. Between September of 1982 and 1983, sixty-eight commercial fishing boats sank off the shores of Alaska. Thirty-six crewmen were killed. In February of 1983 the sister ships *Americus* and *Altair*, the state-of-the-art crabbers out of Anacortes, mysteriously sank outside of Dutch Harbor. All fourteen men were lost, the boats were never recovered, and the cause never fully determined. Between 1980 and 1988, an average of thirty-one Alaska fishermen died each year. Of all the fisheries, crab fishing had the most deaths.

The Derby was partly to blame. Its purpose was to limit the amount of crab harvested, in a fair manner open to all fishermen. First come, first served. The best fishermen with the fastest crews and the best luck got the most crab, but the law of unintended consequences kicked in.

"The competition got greater and greater," remembers Ole Hendricks. "You had to fish regardless of the weather, if you wanted to catch your share of the crab, no matter if it was blowing fifty or sixty. That made it more risky."

Captains saw the advantage to working around the clock. They pushed their crew past their breaking points and chose to work in dangerous storms instead of waiting them out. There were a lot of ding-a-lings up there in those days who hoped to make a quick buck without really learning the ropes.

"There's been as much sorrow as there has been profit come out of the Bering Sea," said Bart Eaton. "If you're lucky like I was, you find your destiny. If you're unlucky, you find your fate."

I entered the crab business just as it was peaking, but almost as soon as the boom started, it busted. In the fall of 1981, no one could find a crab. The King Crab Derby tally dropped 80 percent to 28 million pounds. The next year was even worse: only 10 million pounds. In 1983, the King Crab Derby in Dutch Harbor was canceled altogether, and the Kodiak fishery was closed permanently. The Derby was canceled again the following year.

The scarcity of crab caused the price to more than double, from ninety cents in 1980 to about two dollars in 1981, so a good deal of money could still be made. But by 1985, the industry had tanked, with earnings down to $33 million, the lowest since 1972. The total haul was the lowest since 1958.

It was tough to make a living. The crash of the fishery was compounded by the overcapitalization of the fleet. The megacrabbers built with the easy credit of the late seventies now had empty crab tanks and staggering mortgages. The spike in accidents sent insurance rates through the roof. The cost of diesel fuel had gone up from 18 cents a gallon to $1.25. Too expensive to operate, boats were left moored in Dutch Harbor, Kodiak, Ballard. Some were sold for pennies on the dollar. Owners declared bankruptcy as they defaulted on their loans. Marco launched the last four of their legendary crabbers in 1980, then never built another. The company eventually moved its shipyard to South America.

In 1981 the *Aleutian Monarch*, my dad's floating processor, caught fire and was destroyed. Its charred hull was hauled to deep water, bombarded by the Coast Guard, and scuttled. At the time it was probably a blessing, though, as Sverre's Aleutian Island Seafoods was foundering. The partners sold off the *Xanadu* and shuttered the company.

No longer could you just go up to Dutch for king crab season and make a fortune. You had to fish all the supplemental seasons and fisheries as well. After the collapse, the *Northwestern* stayed busy fishing tanner crab. In the early eighties, it was one of the first boats to fish opilios. The young guys like myself weren't earning full shares during the golden era, so we didn't feel like we were missing out. We were still making money and didn't care if the old-timers told us the boom was over, or that fishing opilios was degrading, or that we were wasting our time. We found ways to make it work.

Opilios turned out to be the bread and butter for the entire fleet. The king crab fishery never rebounded to its 1980 peak, hovering below 30 million pounds for the next twenty-five years, less than 15 percent of its glory days. In the early 1990s, opilios surpassed king crab in tonnage and profit. The *Northwestern* stayed busy all year long fishing opilio crab January through July, blue crab in the Pribilof Islands in August, red king crab in Dutch Harbor in September, and brown crab out west near Adak in November and December.

So while some owners were struggling, others were thriving. In 1985, Dad and a partner, Sigmund Andreasson, bought a 135-foot crab vessel, the *Enterprise,* when its owner couldn't make the payments. It was considered a bad-luck boat, but as soon as Sigmund took the helm, it turned to gold. They overhauled the boat, made a fortune on opilios, then sold it a few years later at a profit.

In 1987, Dad invested that money to have the *Northwestern* lengthened to 118 feet in order to pack more crab and carry more pots. We increased from 156 to 200 pots. In 1991, a pot limit was introduced by

My father bought the *Enterprise* in 1985. *(Courtesy of the Hasen Family)*

Fish and Game. We lengthened the boat to 125 feet to attain the maximum of 250 pots.

The issue of fisherman safety could no longer be ignored. Since 1980, the Coast Guard had recommended that fishing vessels carry survival gear and EPIRBs, but they had no authority to enforce this. Most fishermen and captains—even those who carried such equipment—resisted the regulations on principle. They wanted to be responsible for their own risks. However, with opilios overtaking king crab as the largest harvest, more boats were fishing the Bering Sea in winter, instead of fall. This introduced a new danger: ice on the sea. If boats ventured too far north in the dead of winter, they could get trapped in an ice floe that might crush and sink them.

The accidents and deaths continued. In 1988, the U.S. Congress passed the Commercial Fishing Vessel Safety Act. The bill empowered the Coast Guard to require all crab boats to carry survival suits, life rafts, and automatic free-float EPIRBs that would activate if the boat sank, even if the crew didn't have time to do so manually.

Even these laws didn't make it safe. From 1991 to 1996, crab fishermen died at a rate of 365 per 100,000 workers per year. That's 52 times the national average.

The 1981 collapse of king crab has never been fully explained. Some say that it was caused by overfishing in the boom years, and I tend to agree. But in recent years, overfishing has ended. In fact, after we dramatically reduced the number of males being caught, scientists learned that all the females were still getting fertilized. Half the males did the whole job. In other words, catching the males was not reducing the fishery. However, since no one really knows the natural levels of crab, we can't be sure if the fishery has totally rebounded. Some speculate that changes in water temperature, and reductions in the microorganisms that crab larvae feed on, have reduced the growth rate of the species. I believe the problem now is predation: codfish, halibut, and sole thrive on crab, and their numbers are abundant. It's a natural cycle.

Each year Alaska Fish and Game goes out in the summer and surveys, trawls for crab, figures out how many are in a section, multiplies, divides, and comes up with a figure, and that is our quota. Since I've been a kid I have felt like a guinea pig with a new rule applied almost every year. These days, fishermen work together with the agencies to ensure a healthy fishery. We realize that, if overfished, there is no future for us, either.

Between the collapse of the fishery, the cutthroat competition, the spiraling death count, the tightening of government regulations, and the indignity of having to fish opilios in the dead of winter, it was looking like a good time for some of the old-timers to step back. Dad turned fifty in 1988, and began hiring other skippers to take the wheel so he could spend less time in the pilot house. In 1991, Uncle Karl sold the *Ocean Spray* and got out of the crab business. Three years later she sank.

I got a couple seasons under my belt with the Old Man, but not many. By the time I was on the *Northwestern*, he wasn't fishing much. "Maybe he was a little nervous," said Oddvar Medhaug about Dad's drift toward retirement after the sinking of both *Foremost*s. Maybe he was just tired of it. He'd still go up for king crab, convinced that opilios were beneath him.

When he did come up, the crew called it his "commercial sport-fishing." By then he was something of a legend in Dutch Harbor and Akutan. Mark Peterson remembers pushing through the doors of the Elbow Room with him, and the band cut whatever it was playing and struck up an old Hank Snow song, "I've Been Everywhere."

Sverre had mellowed out a bit since his caveman days. He would take the boat way out west for a short season to Adak or Attu, where there were no lights or boats on the horizon. It reminded him of the early days when he would head out to the middle of nowhere, beyond the reach of civilization, where maybe you'd find crab and maybe you wouldn't, but at least you'd have an adventure. Sometimes their luck was miserable.

"The fishing was so poor that we gave each crab a name," remembers Mark Peterson. That was enough for Sverre. He'd come running out of the wheelhouse, eyes bugging, so excited to have caught even a couple. A black cod in the crab pot was cause for celebration. "He loved them, loved them, loved them," says Peterson. Sverre would salt the cod and eat it; and of course he still loved to eat his not-for-human-consumption bait fish. "It was so nasty!" says Peterson. "It just stank to high hell."

Dad peeling potatoes in the galley of the *Northwestern. (Courtesy of the Hansen Family)*

Whenever Sverre led his young crew on some unprofitable boon-doggle, he'd pay them a decent settlement regardless of the catch, which was almost unheard of in the crab fleet. On one such awful trip Peterson found Sverre up before dawn playing solitaire in the galley, and they made a wager on whether he would win the hand. "If you lose," said Peterson, "we go home today." Sverre played the hand. He lost.

"Go fire up," Sverre said. "We're going home."

And they did.

One Thanksgiving when Chris Aris was onboard, he asked if they were going to get a turkey. "Turkey?" said Sverre. "Ha! You're a tur-key." Later they pulled a pot with a couple of codfish in it. Sverre burst out of the wheelhouse and called, "Hey, Chris, there's your turkey! Norwegian turkey!" Sure enough, they boiled the cod and made a Thanksgiving feast.

Deckhands remember Sverre as shy and surprisingly compassionate for a skipper. "He was just like an uncle to us," remembers Mangor Ferkingstad. "Fair and square. Always looking after us."

Late one night Chris Aris had an earache and was rifling through the medicine chest.

"You got eardrops?" he asked Sverre.

"Yeah, there should be something in there."

Chris squeezed some drops in his ear and crashed in his bunk, knowing he'd be up again in a few hours. He slept like a rock. When he awoke, eight hours had passed. *Oh shit*, he thought. He hurried into his raingear, ran out on deck, and apologized to the crew. "Don't worry," they said. "The Old Man said to just let you sleep."

"I'll always remember that," Chris says. "He did something really nice when he didn't need to."

Sverre didn't come out on deck much. The wheelhouse was his castle. Sequestered up there, he filled one coffee cup after the next and snuffed cigarettes into a heap in the ashtray. At night, he would listen to the KMI radio channels, where you could hear fishermen on other boats calling home. Guys talked to their girlfriends or their wives. It was like listening to a soap opera.

"I love you so much, baby," said the girlfriend, her voice crackling with static on the radio.

"You better watch out, you dummy!" chuckled Sverre. "Next thing you know she asks for money."

"I love you, too, babe," came the fisherman's distant reply.

"Can you get another draw and send a little more," said the girl-friend.

Sverre shook his head and laughed. "Don't do it, you dummy!"

"Things are a bit tight right now," said the fisherman.

"Oh, baby," said the girlfriend. The line was silent.

"He's going to buckle," Sverre announced. "I betcha he buckles."

"I'll send the money," said the fisherman.

"What's the matter with this guy?" cried Sverre. "The dummy!"

Sverre also liked to have some fun. "No matter how bad a situation got," remembers deckhand Tim Canny, "Sverre could always laugh about it." When Mark Peterson was a greenhorn, Sverre looked up from his dinner in a panic and said, "Did anybody feed the crabs?" Everybody turned to Peterson.

"What do you mean?" said Peterson.

"Bait boy is supposed to feed the crabs," Sverre said. "You didn't feed the damn crabs?"

There were a hundred thousand pounds of live crab in the holds.

"I didn't know," Peterson faltered.

"It's been seven days! How come you didn't feed the crabs?"

Peterson shot up from the table and ran out on deck and ground up fifty pounds of bait and dumped it in the tank. He did that for four days. Finally Norman pulled him aside and said, "You know, Mark, you don't really need to feed the crabs. He's just messing with you."

Another time Peterson was up in the wheelhouse with Sverre when they were anchored off Attu, far west in the Aleutians. There was nothing on the island but an old loran station operated by the Coast Guard. The World War II docks were so old and rotten that to tie up you had to wrap the line around the entire dock, because any small part of it might just bust off. It was a cool place to explore. There was an old airplane, ships that had sunk in the harbor, and little ammunition huts where men had scrawled their names on the inside walls. If you ever broke down in Attu, you had to hire a plane to haul parts and mechanics from Dutch Harbor into the old Coast Guard landing strip.

Then the weather went bad. The wind was blowing hard. They had a full load of gear on deck, so they were unstable. "We always knew when the wind got over a hundred," says Peterson, "because the radar would stop turning and the screen went green." They pulled anchor to seek shelter on the lee side of the island.

They were at the spot where the Bering Sea and the Pacific Ocean meet, exposed to the full force of both. As they came around the point, Peterson on portside and Sverre on starboard, Mark saw waves like he'd never seen. "I'm looking out at the seas, and they look like— you know those old pictures where you feel like you can see through the water forever and there's veins and foam and stuff? We're in seas like that, and I don't know how tall, and I'm not even going to try and guess. Biggest things I've ever seen. Huge."

They ran with the seas, rather than buck into them, and the monster waves crested behind them and crashed over the stern of the ship. "All of a sudden the boat just shears off to one side," says Peterson. "We started rolling up on our side, and the propeller started sucking air so all the alarms were going off and the main engine was sitting there without water circulation and the bearings were overheating. And I could see water outside of Sverre's window. He's sitting there, and I was hanging onto my chair, and my feet were hanging off. If I'd let go, I would have fallen all the way to the other side of the wheelhouse."

The boat rolled completely on its side, and Mark thought they were going over. "Somehow, the boat righted itself. I don't know if he hit the jogstick. I don't know what happened. But we came out of it. I remember going downstairs and looking around and every single thing in that boat was strewn everywhere. All the doors were open. All the guys were sitting there, and nobody really said anything, looking at each other like, what just happened? Unbelievable."

Sverre hardly reacted. "He didn't jump up and high-five and freak out like we did." He just lit another cigarette and drove the boat to the cove.

Another time Mark Peterson got the idea to have jackets made for the crew. Nowadays most of the boats have crew coats, but back then it was rare. Peterson led the charge. "I wanted the jackets so bad I could taste it," he says. At the time, the only jackets people were making had a little logo on the chest. Mark wanted the biggest thing they could put on the back of the coat: a huge drawing of the ship itself. He went down to Custom Embroidery in Ballard and got all the information and put it together. Everybody wanted coats. All he needed was a really good picture of the boat for the design. He dropped by the house and told Sverre what he needed.

"What do you want that for?" Sverre barked.

"Well, I want to get some jackets made."

"Jackets!" sniffed the Old Man, like it was the dumbest thing he'd ever heard from this bunch of silly kids. "What do you want jackets for?"

"I know, I know," said Peterson. "But *we* want them and *we're* going to pay for them. I just want to know if you have a good picture."

So he gave Mark the photo, and Mark designed the jackets, with the picture of the boat and the words F/V *Northwestern Seattle Washington*. When he was done he returned the photo to Sverre.

"What color are they going to be?" Sverre barked.

"Blue."

"Blue?" scoffed the Old Man. "What else?"

"I'm going to put red and black on the back."

"Red?" he cried. "Dummies!"

"Whatever," said Mark as he, turned to leave. "I'm heading this up because nobody else wants to."

Then Sverre said, "Well, better get one for my father and me."

Of course, Dad ended up paying for everyone's jackets. "It's funny," says Peterson, "because I kind of knew he was going to do that all along, but I didn't want to assume."

Sverre was fair, he paid on time, and he was willing to let a crewman draw on his settlement early. When Matt Bradley called to ask for a holiday advance, Dad called him over to the house for a cup of coffee. Matt was battling his addiction. The Old Man laid two envelopes on the desk. "One has five hundred, the other has five thousand," said Sverre. "Pick one."

"I don't need five grand," said Matt. "Just enough to buy Christmas presents."

"Pick one," said Sverre. Then he smiled. "But if you pick the five thousand and don't make the boat on time, I'm gonna kill you."

Matt chose an envelope and tore the seal. Five thousand dollars.

"I knew you'd pick that one!" Sverre cried.

The two of them were good friends. Even when Matt went through hard times, the Old Man stuck with him. He saw his own fighting spirit in this underdog.

"Sverre was more of a father to me than my own father," says Matt.

When my cousin Stan was working on the *Northwestern,* he would ask Dad for an advance at the beginning of the season. Sverre said, "How much you want? Five, ten, fifteen thousand?" Stan would have been embarrassed to ask his own father, so Sverre kept it a secret. On that infamous trip where Steiner almost got pulled to his death and we didn't make any money, Sverre paid him anyway. Another time Peterson quit the boat, and then asked for his job back. Sverre had already hired a replacement, but he rehired Peterson as a fifth hand, and instead of reducing each man's share, he paid Peterson from the boat's share. During a St. Matthew season Sverre promised the crew seven percent, then gave them nine.

Most of the crew was young, and didn't understand how to save money for tax time. In bad years Sverre paid their taxes up front, then deducted the amount from later checks. Once he even cosigned for Mangor Ferkingstad on a new car. Mangor remembers, "Pay was good, gear was good, the boat was spic-and-span, and the boat almost always caught a lot of crab." Most guys kept their jobs for years.

"We were like soldiers for him," says Mangor. "We were proud to be on the *Northwestern*."

9

LIGHTS, CAMERA, FISHING

*C*aptain Sverre felt his legs stiffening as his trousers froze solid. They had drifted for three hours. There wasn't much to say and not much to be afraid of, either. He had that Scandinavian stoicism deep in his blood. They all did. It didn't matter if there was any chance of survival: complaining or being afraid didn't help. All you could do was wait. Sverre was willing to hope, but not if doing so took any energy. He needed his strength just to continue flexing his knees and elbows so they wouldn't lock up. He was cold, colder than he'd ever been in sixteen years on northern seas. He trembled in his wet sweater. The others were shivering uncontrollably. Every half hour or so they opened the flap and looked out to the sea. Nothing.

They had plenty of time to think. Memories passed through their minds. There was the time Leif Hagen lost his wallet in Alaska and couldn't get a ticket home. He called John Jakobsen, who wired him the money. There was the time Magne Berg had been working with John

Johannessen, and they were tied up at the Alitek processing plant near Kodiak and had been partying onshore in the bunkhouse. John came back to the boat first, and was sitting up in the pilot house when Magne climbed below to sleep. John decided to have some fun. A half hour later, he climbed down below and rousted Magne. "Better come and steer," he said. "It's your watch." Magne dutifully awoke from his drunken sleep and hauled himself upstairs. He sat in the captain's chair and looked out into the black night and kept the boat straight. John sat beside him for a while and watched, trying not to laugh. It took a few minutes before Magne's eyes came into proper focus. Then he looked carefully to the starboard and saw that the boat was still lying in the dock.

Another time the guys were playing poker and Krist Leknes was losing bad, so he borrowed two hundred from Karl Johan. He didn't pay it back. One day Karl found him at Vasa, his money scattered all over the bar. "Boy, you better get your money together," said Karl. "Ah, never mind," said Krist. "I owe you two hundred, but you don't need it anyway." The next time they met, Karl let him have it. Boy, did Krist get to hear about it that time. In the morning Krist knocked on Karl's door and paid up.

These were the stories the Karmoy boys repeated around the warmth of bars and galley tables, stories that always brought a laugh and a round of heckling. Now stranded on this wet life raft, the memories were bleak and bitter.

"Well, would I love to hear the roar of an engine right about now," Sverre said. The others managed a grin, and returned to their silent misery, curled up against each other in a pile on the floor.

———

We're approaching the end of my father's saga—his upbringing in war-torn Norway, his immigration to Ballard, and all his hard work and good fortune that allowed me to jump into the fishing industry at a young age. I have him to thank for my own successes. Now you know how I came to be skipper of a crab vessel, which up until five years ago would have been the whole story.

Then the strangest thing happened. I went on TV, and people started to recognize me. I never could have predicted it. Like anyone who goes on TV and gets even slightly well known, I like to think that it hasn't changed me in the least, that I'm still the same hard-working, home-loving guy I always was. I try not to let it go to my head. I understand that fame only lasts fifteen minutes, and that next season the public will have found some new action figure. I get it.

The truth, however, is that all the publicity and opportunities that have come with the *Deadliest Catch* have affected us. I'd be lying to say otherwise. Take this book, for instance: Five years ago I wouldn't have thought to write it, and nobody would have thought to read it. That's just the truth. So I want to talk about what it's like—how crazy it really is—to go almost overnight from a lifelong blue-collar fisherman to a guest on the Leno show.

Fifteen years ago, long before the TV show, I was pushing my cart through the grocery story looking at the bottles of Paul Newman salad dressing on the shelf. It hit me: This is a real person on a bottle of food. So I pushed my cart over to the seafood aisle. What did I find?

Charlie the Tuna. A cartoon. Star-Kist had a mermaid. And then there was Gorton's—some guy with a yellow hat. No actual human being appeared on any of the boxes or cans.

So I started thinking—Why not make our own brand of seafood? I wanted more than just crab. I wanted fish sticks, salmon, cod filets, halibut steaks with an ink-stamped image of the boat. The back of the box could have the history of the *Northwestern,* and tell about how the Hansen family caught the different types of fish. Most people don't know how their seafood is caught, and it would be cool to tell them. Every product has a story.

I floated the idea past some retail chains, but since I didn't have control over my product after it went to the processors, and because I didn't have enough year-round volume to keep the product on the shelves, they thought it wouldn't work. The idea died.

Seven years later, Edgar and I got word that a producer wanted to make a documentary about crab fishing. His name was Thom Beers, and he'd gone up to Alaska and made a trip on a boat called the *Fierce Allegiance.* He thought it was just amazing, so he came back to Seattle and interviewed all the captains he could find. That's when Sten Skaar, who runs the *North American,* mentioned us. "What about the Hansen brothers—another Norwegian family?"

We were the last guys to interview. We set up a time to do a Q&A on camera, but the producers were late. Edgar and I were waiting around in this Chinese restaurant and had a few drinks. They still didn't show up. A couple hours passed, and by the time they arrived, we were half in the bag. I guess that took away our fears of being on camera and made us more natural when the cameras rolled. We started pecking at each other in good humor, the way we often do, and apparently they liked the brotherly "love."

At first I didn't want to do the show, but then I started thinking: What the hell? It's a documentary, a onetime deal. We can get our family in there to present our legacy. Our family has been in the industry since Day One, and we take a lot of pride in what we do. We thought we could depict the industry in a positive light. So why not? The show was going to happen anyway, whether it was us or another

A stationary *Deadliest Catch* camera mounted to film the deck, covered in ice. *(Courtesy of EVOL)*

boat. We accepted their offer to do a three-part documentary and allowed the film crews on our boat.

After the first season aired, the reviews were great, the viewers loved it, and they asked us to do another full season. We realized that we couldn't buy advertising like this—and that it could possibly cause and increase the demand for Alaska crab. So we figured we would pursue it. I thought, *Let's get a Web site. Let's sell some T-shirts. The OC Chopper guys did it. Why not us?* Back then I didn't even own a computer. So I asked my neighbor to help me set up a site at his house. He showed me how to do it. It had a fan box and a comment box.

Next thing you know, *bam!* People Googled our name because they wanted to know about the boat and the industry. They began to visit our site. The e-mails flooded the comment box, mostly just compliments.

People said, "I could never do what you do. We have so much respect for you. I'll never argue about the price of crab on my plate again." Their comments were almost always positive.

When I started to respond personally, the e-mails started coming directly to me. It was fun at first, but I was just trying to progress from one-finger typing to two, and I couldn't keep up. People expected me to do it quickly. I lost count of all the e-mails. We had to boot it up (or whatever you call it) to make the site bigger. I always thought I'd run the business with paper and phone calls the way Dad ran his business, but suddenly I had to join the twenty-first century. It became more than I could handle by myself.

Matt Bradley's brother suggested CafePress, an online clothing company, and we sent in a few designs. One day on the boat I didn't want the crew to bother me. I hung a sign on my door that said, "Shut Up and Fish." Everyone thought it was pretty funny, so we decided to use that as our logo. I told CafePress to print it on whatever they wanted. They stamped the logo on shirts, sweatshirts, hats, and even on women's underwear. I didn't have any idea what was going on. They started selling. We sold thousands of them. When I went on Jimmy Kimmel's show he held up a "Shut Up and Fish" butt thong and said, "What the hell is this?"

The *Deadliest Catch* stirred the pot at Dutch Harbor. Some of the big names and influential people up there turned their backs on us. Literally. When we walked into the bar and said hello they turned away, didn't even acknowledge us. Some people made comments, like "Hey, Hollywood!" or "Here come the famous people." They looked at me as though my crew and I were sellouts, like we were going to hurt the industry. They were paranoid that we were going to sell the fleet down the river. It was already competitive, and this created even more tension. One captain brushed past me in the bar and said, "You don't represent the crab fleet." They thought I was trying to be *the guy*, although I never *asked* to be the guy. One skipper complained because the show made us look like highliners, when actually other boats catch more crab. *Really, you think? Well you're not in the documentary.*

Like fishing, popularity is also competitive. You've got your image on that electronic box and people watch the box, so you might as well make the most of it. There were four to six boats on the show, and people began to pick their favorites. That's just natural. It was heart-wrenching to think that if we had a bum season, we would look like a bunch of jackasses. And there goes any opportunities.

After the second year, we saw a potential to actually capitalize, and maybe make a brand for ourselves. That's America. That's a beautiful thing. It's just like our Old Man and the rest of the fleet used to do: if one boat does well, build another.

Then Edgar and I went to Boston for a seafood convention, and we were the keynote speakers. People showed up! It was the first time that room had ever been packed by a speaker. I also did the keynote address at Fish Expo in Seattle. Same thing. Finally, for the first time, there were well-known fishermen speaking for the industry. In some people's eyes we were like celebrities. Little kids tell their mom they want to dress up like Edgar for Halloween. When Discovery saw the video clips, they said, Those are our guys. That was a stroke of luck. So they started sending us out on talk shows. Jimmy Kimmel, Jon Stewart, Leno, Conan.

We came off all right on those talk shows. We were doing well on the show as professionals, and we gained a lot of respect that way. We didn't take it very seriously, we had fun with it, and people could read that. It didn't really hit me how big this had become until I was down in Las Vegas with my wife. We were watching this guy across the room in one of the hotels. All these girls were getting his autograph.

We knew he was someone, but we didn't know who. So my wife went over and introduced herself. She came back and told me that it was Vince Neil, the singer from Mötley Crüe. I know the music, but I didn't recognize him. My wife said, "Go take a picture with him." Vince Neil was thrilled. "*Sig,* dude! I love your show." I was thinking, *Isn't it supposed to be the other way around?* He turned out to be such a generous down-to-earth guy. I really liked him. Then his wife said that she didn't like the show. "Every time it's on, he watches rerun after rerun, and he never goes to bed!" The guys and I got a kick out of that. To think that a crew of humble fishermen may have messed up an actual celebrity's romantic life on Tuesday nights.

A few times when we were in L.A. we stayed in the Hotel Roosevelt. I've had bartenders and busboys approach me and say that they've seen a lot of the Hollywood big shots come through those doors, but that they were more impressed by the fishermen on the *Deadliest Catch.* That made me feel pretty good. We're connecting with those workingmen. They know that a movie's a movie, and an actor's just an actor. Come to find out, they're connecting with us on a different level. Most celebrities are all about creating their image, but we're just doing the same thing we've done our whole lives. A lot of guys respect that.

Of course, there's a downside. One time on the show I had a little hissy fit over a crab count. I was freaking out, slamming things on the table. After it aired, I'd meet people on the street, and they would imitate me. It was embarrassing. Half the time I don't even remember what I did on camera, and six months later I'm surprised to see how I come across. Most viewers understand that we're not like Brad Pitt or Tom Cruise. We're just fishermen. That's why most people just say, "Good job," or tell us that they respect the work we do.

So the image of fishermen is changing. You can think of commercial fishing like professional baseball when it first started more than a hundred years ago. Back then, the owners made all the money. The players were lowlifes: hardly paid, no job security, and no benefits if they got hurt. They were replaceable. If one guy squawked for more money, the owner could find a new guy. That's how fishermen have always been: replaceable cogs in the big machine. My dad used to tell about how one of the richest highliners in the fleet tried to buy a house in a private gated neighborhood called the Highlands. The doctors and lawyers who lived there voted him down, even though he was a millionaire. They didn't want fishermen in their enclave. It showed me how there was a pecking order in our society. When I was a young man, if you went to a nightclub and told a woman you were a fisherman, she turned up her nose. We were considered lowly. These days, through the work of groups like the Alaska Crab Coalition, and with the publicity around the show, fishermen across the country get more respect from the general public.

Now up in Dutch Harbor, some people think differently about us. Instead of the rude comments, some are complimenting us for doing a good job and for not coming off like jackasses. Unexpectedly, the show has led us to get involved in some great charity work. We were invited to help with fund-raisers for the Ronald McDonald House charities, the Nordic Heritage Museum, the Children's Hospital of Seattle, and the Fred Hutchinson Cancer Research Center. We raised money for an elementary school in Gloucester, Massachusetts, to build a new playground. At a benefit for his Ally's House charity for kids with cancer, the country singer Toby Keith auctioned off a two-day float on the *Northwestern* for $28,000. The winners joined us during the annual Sea Fair festival in Seattle, where we had been named grand marshals. Also onboard was Gary Yost, a ten-year-old with bone cancer, whose wish through the Make-a-Wish Foundation was to fish for crab on the *Northwestern*. Fish and Game signed a special permit for us to drop pots in Puget Sound, and Gary had a blast helping out on deck. I named him honorary Captain for a Day. It was a great day for us and for Gary, and we still stay in contact with him.

Through this all, I realized that we had become the public face of fishing. Finally I had a shot at marketing my own brand of seafood. I knew the potential was there, I just had to talk the processors into selling the product. At last, after two years of negotiating, I got one of the big processors to cobrand with me. I was thrilled to be a part of it. They stuck our picture on the back of the box, and placed it at Costco. It didn't sell well. Crab is a finicky market.

Then they offered to put my crab in restaurants, but they only wanted what was caught on the *Northwestern*. It would be a specialty item, like Copper River salmon, except that we only catch 300,000 pounds, and it's not enough to go around. Half of it is shell, so we only get fifty to sixty percent recovery. That's only enough for a handful of restaurants for a year—and not many restaurants would be willing to take the risk of buying a whole year's supply of frozen crab all at once. I'd be missing out on an opportunity.

So what else could I do to market my crab? To see how tricky it is to sell seafood in the United States, you have to understand how complicated the industry is. First off, the majority of king crab that Americans catch in Alaska—what you see on the *Deadliest Catch*—ends up in Japan. The Japanese were fishing Alaskan crab long before us, and by the time they were kicked out, they had developed the infrastructure—and the demand—for the product. Much more so than Americans, the Japanese are willing to pay top dollar for the best-quality crab in the world. It's been a staple of their diet for centuries—we eat steaks and burgers;

they eat fish. In fact, Japanese fishermen have long been revered, and haven't had to fight for respect the way we have in this country. The consumers like the uniform size of the Alaskan crab, and as a result, Japanese importers are willing to pay cash up front. In other words, American fishermen and processors don't have to take the risk that if their product doesn't sell in stores, they won't get paid. If it weren't for Japan, we wouldn't have an American crab industry today.

A small percentage of Alaskan crab stays in the United States, but you can usually only get it at fancy restaurants or gourmet boutiques. It can cost a pretty penny. So where does the crab in your local grocery store come from?

Russia.

There's king crab there, lots of it, but the Russians have their own set of rules. They don't follow the same strict size limits that we have in America, and, of course, they don't have to follow the quota or Coast Guard requirements that we have here. As a result, they bring in a lot of crab that is smaller than what we catch in Alaska. Ironically it was the American fleet that went over to Russia in the 1980s and taught them how to fish, and sold them a bunch of boats. Now the Russians are underselling us with smaller product. It's a problem of our own making.

Every year, the Russians export hundreds of thousands of pounds of frozen crab to the world. If it doesn't sell over the course of the year, distributors are stuck with a surplus. At the end of the season they dump all that crab on the market at Albertson's or wherever, for nothing. The price drops. People wonder why they should pay twenty dollars a pound at a restaurant when they can get it for a third of that at Albertson's. When the price drops like that at the cash register, the processors in Alaska justify cutting the price that they pay us fishermen. Never mind that our crab ends up in a different country—they cut the price anyway.

Here's the point: For years I went to these big American seafood companies and told them to put their crab in a box with my picture and sell it. They didn't want to. The American stores would only pay them after the product had already been bought in stores. It was easier for them to sell it to Japan, and they got paid up front. To do it myself I would have had to buy from them the crab I just caught at a markup, mark it up again in my box, and then sell it, with me taking all the risk. I wasn't going to do that.

That's when the Russians approached me. They're actually an American company that imports the Russian product, the number-one importer of crab in the United States. They saw me speak at Seafood Expo, and they knew they could make it happen. They told me that they had done the Jimmy Buffett Margaritaville Cafes deal. It was huge. So they said to me, "You are it."

We had a meeting and we knew it was going to get sensitive. They knew I would probably take a hit, putting my name on Russian crab. So these guys told me they would not dump all that crab surplus on the market as they had done in the past. Instead they put it in a box that said "Captain Sig's Northwestern," with our picture on the front, and sold it at Walmart for nine bucks a box. It was a huge chance to build a brand name, so I went for it.

Turned out that physically getting crab legs into the store was not so easy. They had to design a box that would not only hold exactly two pounds of hard-frozen crab legs, but also nestled perfectly into a container that fit on a freight truck, and also matched the shelf size at Walmart. Most fishermen don't know that—I just learned it recently. To be quite honest, I'm amazed by how much I didn't know about the complicated nature of our industry.

The box looked great. It had a picture of the boat crashing through a huge wave. One of the account reps, who'd never heard of the *North-*

western or the show, said, "This is the coolest box I've ever seen. I want it in my stores."

The product flew off the shelves. Instead of these legs going into Albertson's and knocking down the price, they went into my box. It got the surplus of those smaller legs out of the marketplace. Once they were out of the market, the demand increased; the price actually went up seventy cents a pound in stores. That increase directly affected the price we were paid by processors in Dutch Harbor. I'm not saying that my box of Walmart crab legs caused the price to go up, but it helped stabilize the price, and didn't cause it to go down.

I took a hit for it, though. The fleet wanted to cut my head off. We had to put "Product of Russia" on the back of the box. What we didn't mention was that it was a Russian product that was in the stores anyway. If they had dumped their crab on the market prior to us fishing that season, it would have been disastrous. Most of the flak, though, was internal, from the fleet. Even some of the fishing gazettes, which have a reputation for being critical, ended up supporting me on this. They never really wanted to admit it was a good thing.

It didn't help my case when the president of the Russian company got sideways of Vladimir Putin's government—the company was catching too many crab and making too much money, and the government wanted a piece. In recent years the government has taken control over more of the natural resources. So that deal ended after one year.

Our product actually opened new doors for my family. After it succeeded, the American processors came back to me. These executives took a trip to Walmart, and they loved what they saw. So we negotiated. By the time the deal was complete, the box had changed a bit. They thought *Captain Sig* was a more recognizable brand than *Northwestern*. They also designed a box with colors that would stand out against other brands already in the Walmart freezer aisle. In 2009, the new products hit the shelves of three hundred stores for a test run. Captain Sig Battered Pollock, Battered Salmon, Salmon Burger, Popcorn Fish, Fish Sticks, and Hidden Treasures. For now, there's no crab because the premium product would cost fifteen bucks a box, and they figure that wouldn't fly at Walmart. We'll see how our new products do. It's just like fishing: You throw something out there and see what you catch.

We've got other products out there, too. A line of Tastiest Catch sauces: tartar, cocktail, horseradish, and Crab Louie. Rogue Brewery in Oregon makes Captain Sig's Northwestern Ale, Helly Hansen sells *Northwestern* rain jackets, and, of course, there's the Xbox video game. I'm happy to let these partner companies do the work, and I'll just give the endorsement. I've heard horror stories about sports figures and celebrities. Rather than taking the percentage and letting the other guys do the work, they try to do their own business, and they fail miserably. You've got to know your place. At this point, none of these products are making a lot of money. We're still fishermen, and that's our bread and butter.

I've got other ideas. Why not do a restaurant promotion? I'd like to do a thing where you order king crab legs, and then they come on a plate and if you turn one over and there's a *Northwestern* logo stamped on it, it means it came off my boat, and you win a prize.

It's important to expand our business, because I don't know how much longer the family will be fishing. I have a feeling that my brothers and I will be the end. The way that we've worked, the way that we've carried our quotas and maintained the integrity of our family—I think it's finished. I don't think our kids are going to want to do it. They're just too modern. To them fishing is barbaric. Times change.

Every ten years there's a new generation. Guys eight years older than me are from a different culture. Now that it's more of a business than just a fishing boat, the kids don't have to be on the boat. The boat can still make money and provide for them.

So that's how Captain Sig seafood ended up on the shelves of stores. That was my goal for fifteen years, and we did it. We realized that all these efforts may come and go, but the boat is our permanent livelihood. So we still continue fishing.

By 1990, the new generation was taking over for our dads. Although we were thoroughly Norwegian, we were all born and raised in Seattle. We had our own lives. As we turned twenty-one, we had little interest in driving down to Ballard just to have a beer with a bunch of old-timers. Besides, Ballard didn't belong to us the way it had belonged to our fathers. A new crowd was moving in, not the working class that had been there for a century, but white-collar professionals. JCPenney's was closed, subdivided into chain stores. The old houses were bulldozed and replaced with condos.

"The old-timers kind of lived down at the Smoke Shop," remembers Chris Aris. And if you had to find one of them for something, and it was already three o'clock, forget it. Because these guys had already had enough early on in the day. You go in and everybody's going to buy you a drink. Well, what are you going to do, turn around and say, 'No, I'm not going to drink?' Next thing you know you got drinks stacked up. And I kinda got things to do here. I don't want it to be eleven thirty in the morning and have a couple of real strong drinks in me. But what are you going to do? So I really didn't go down there too much."

"You didn't spend too much time down there because they started questioning you about everything you did," says Lloyd Johannessen.

A wave crashes over the rail on an icy opilio trip. *(Courtesy of EVOL)*

"They start giving you a hard time and talking about how tough things used to be, and how kids today are idiots. They talked about rocks and burlap sacks and what they should have done with their kids."

The new generation found our own place closer to home, and the heart of Norse crab fishing moved north 150 blocks from Ballard to the North End. We settled into Duffy's, a local watering hole owned by a crusty old Irishman named Duffy. Duffy loved the Norwegian fishermen—he even hung a sign over the building that said NORWE-GIAN CHAMPAGNE SERVED HERE. That was a drink that the crowd my age had adopted as our favorite—vodka Coke. Duffy was an old cod-ger, a onetime boxer, who didn't check ID. He'd arm-wrestle men twice his size, and half the time he'd win, flexing his bicep like a Pop-eye muscle. He was a great fighter, and you had to watch out or he'd head-butt you.

We'd been seen sneaking into Duffy's since we were nineteen years old. Sometimes he'd take money right out of the till and go down to the track to bet on horses. He'd leave me in charge of the bar and the till. If you asked for a beer, I'd give you a vodka Coke. There was a button on the tap that said VC, and that was the one I pushed—no matter what you ordered.

Duffy protected us. If he thought there was an undercover cop in the bar, he'd stall him until we got outside. Looking back, I suppose Duffy was to my generation what Inky Boe was to my dad's—the bar-keeper who watched over us and kept us in line.

Eventually Duffy sold the place. The bar's new owners, Marsha and Robin Stiff, had moved from Montana hoping to buy a bar. They looked at fifty bars. Marsha thought most of them were too froufrou, "fern bars waiting to happen." But they liked Duffy's, with its crowd of locals and a few pool tables. When they took over, they brought in dartboards, a huge jukebox, and a shuffleboard table. We young fisher-men pretty much took over the place when we were in town. Anytime one of us dropped in, it seemed there was another fisherman hanging out to talk to. There were times when every single bar stool was oc-cupied by a Norwegian. It was the kind of place where you could buy a shot of Maalox. The Stiffs renamed the bar the Wild Horse.

Marsha would take fishermen to the airport if they needed a ride.

Some of the guys called her Mom. A few even dated her daughter, and Marsha referred to them as "my future ex-sons-in-law."

Marsha remembers, "They were a bunch of great, obnoxious kids that I fell in love with. And I was the mom. If they got drunk, I'd take them home. If they did something wrong, I'd yell at them. If I caught them in the women's room, making out with some broad, I'd say, 'What the hell are you doing?' "

One guy was always broke no matter how much he made. So Marsha banked his paychecks for him. Then she'd dole out the money when he needed it. "Fifty bucks will get you shitfaced," she'd tell him. When he wanted to go back to Norway for vacation, he asked for a thousand bucks. Marsha said sure. He looked at her funny. "What do you mean, sure?" he asked.

"Well, you have $4500 in the bank!" said Marsha.

He couldn't believe it. "This is way cool," he said.

Not a lot of single women hung out at the Wild Horse. "I don't think American women got it," says Marsha. "Remember, they weren't famous or anything; they were just a bunch of knuckleheads that liked to go fishing. American women wanted a boyfriend around all the time, and these guys were gone. You have to worry about them all the time. Accidents happen way too fast. Nothing you can do about it."

When a fisherman did find a girlfriend, he brought her to the Wild Horse to meet Mom. That's when you knew it was serious.

Marsha was there every morning at eight, but didn't open the doors until ten. "But everybody knew, if my car was there, guess what? They'd come in the back door and help me vacuum and set up."

Of course, she did her share of mothering. "They were good kids," Marsha says. "They knew right from wrong. They just needed someone to hit them over the head. When he starts drinking, Sig's voice gets louder. You just hear it. He gets so excited. I had this Nerf baseball bat in my office, and I'd come out and just hit him on the head with it when I had to."

Marsha remembers the first time I met with the elders of an advocacy group, the Alaska Crab Coalition, at the Wild Horse. Marsha remembers, "Sig had an opinion. Everybody listened to the older guys, but Sig thought the younger guys had something to say. So we sat there one day at the table and went over what he wanted to say. And I said, don't get mad. To make your point, you have to calmly tell the president, that you have something to say that's worth hearing."

We sat down at a table and had a couple drinks.

"I could see that Sig was getting upset. He felt he wasn't getting his point across. I thought, *Oh, here goes the voice.* But he kept it together and made his point. After that, the young guys started getting involved. At that point, they knew they were ready. It had become their own. It was time to move on. Let the dads know, we're ready, we're not just the kids anymore."

Eventually the old-timers followed their sons and adopted the Wild Horse as their own. On occasion some of the old guys even brought their wives. It became a family bar. No longer was there a divide. They dubbed it in Norwegian the *Vildehest.*

Now in their fifties, some of the old-timers decided to slow down. Instead of driving to Ballard every day, they would stay home in the suburbs and play croquet. It was the mafia all over again. There was a problem, though. Even though they were old, they were still competitive! A few rounds into croquet, there was fist-pumping and trash-talking. Next thing you knew they were practically swinging mallets at each other. After some croquet parties they wouldn't speak for weeks. My Old Man got a laugh when he heard these stories down at the coffee shop.

Spending less time on the boat, Sverre had more time for his other passions. He traveled to Bergen, Norway, for a soccer tournament and took time to see the statue of Shetlands Larsen, which he deemed too small. "They should redo it," he griped. "Make it twice as big!"

When Edgar and I were working together on the *Northwestern*, Dad came up to Dutch Harbor for king crab season as sort of a last hurrah. It was the only time the three of us fished together as adults. Dad said he was just coming as an "observer." At night he was down in the galley helping cook, but the rest of the time he was with me in the wheelhouse.

"Don't worry about me," he said. "I'm not here. Just go set the gear."

Fine. I went to do my thing and set the gear. But if you've ever tried to do your job with your dad looking over your shoulder—especially if he's a master of the craft, and your mentor—you know it's damn near impossible. He didn't criticize me, but I was so uneasy. I was freaking out. My gut instinct was to set a course east, so I did, but then I saw a couple boats heading west. Uh-oh. Wrong way. I always second-guess myself like that, and usually my last guess is the right one—it's what's in my gut that I ignore. Then my dad piped up.

"You can't change your mind just because of what other people are doing," said the Old Man. "Just make up your mind and do it."

Then I felt I couldn't change course because I'd look wishy-washy. So I continued east, and we dropped the pots. All the while I thought, *I should have turned around and gone west.*

We checked three pots out of a thirty-pot string. Every single pot had crab in it on an early soak. It was good indication. We started pulling more pots, but they were blanks, nothing in them. The only crab in the whole area was in those three that I checked. I was annoyed. I knew it was because I'd been pressured by my dad, and hadn't made my own decision.

We were all frustrated. The Old Man said, "Go to bed, go to bed." I didn't want to go to bed, though. I wanted to catch some crab. The Old Man insisted, like he was doing me a favor by watching the wheel while I slept, but that wasn't it at all. As soon as I lay down in my stateroom, I heard the crane, the power block, and the crew working. They weren't sleeping. Dad just wanted to pull a few pots, for old times' sake.

Like most Norwegians, Dad was crazy about keeping the boat clean. Even though he wasn't skipper, he couldn't resist giving us some pointers on how to do things his way. It was his boat, after all, but none of us wanted to hear it. One day during some very rough seas he sat in the portside captain's chair while I pulled pots. Suddenly we heard this crash down in the galley. I ran down and had a look.

Every single dish had crashed off the rack, hit the wall, and shattered on the floor. I started screaming. "Fucking shit, those goddamn idiots! How can they be so stupid to not put those dishes away." I called the crew in off the deck and ranted and raved. "If you're going to take the time to wash them, take the time to put them away. You're on a freaking *boat*." So the guys started blaming each other. Edgar accused Matt of leaving the dishes out. Matt denied it. Everyone was pissed. I looked at the Old Man over in the captain's chair. He pulled his hat down, sank down in the chair, looked out the other window and muttered something under his breath.

I didn't learn the truth until a couple of years later when I was in Norway and ran into one of my dad's best friends. "You never learned the truth about those dishes?" he asked. "Me and your old man laughed for hours about that one night."

It was Dad who left the dishes out, and he never told me. Of course, his friend had to tell me, in Norway. Even when I brought it up to him, he just laughed. The Big Kahuna sinking down in his chair!

Although Sverre was no longer running the boat, he still liked to be a part of the team. Whenever the crew was flying up to Alaska, we all met at Dad's house at five o'clock in the morning. He just wanted to see us off, and make sure everybody showed up. Mom would give us hugs. "OK, bye—don't do anything stupid," she'd say. After he gave us his two cents we'd carpool to the airport.

"I can honestly say I'm the only one to hug my dad," says Edgar. "Yep, every time I left for Alaska. It was weird. I started realizing, shit, this might be the last time I'm coming home or leaving. I'd go to the bathroom and wait until everybody was in the car. Then it was just me and the Old Man.

" 'Yeah,' he'd say. 'Okay.' "

" 'Yeah,' I'd say. 'See you when I get home.' He'd still be standing there by the front door, stiff as a board, hands in his pockets. I'd just give him a little pat. Then he'd get this look like he'd never been hugged before. 'Okay, have a good season. Be careful. Don't break my boat.' "

This went on for a couple of years. "I was about to give up on it," Edgar says. He wondered if the affection was being wasted on Dad. The news of Edgar's hugs made it through the grapevine to the Wild Horse. That's the way our family goes—you will never hear anything positive, unless it's from somebody else. Edgar was down having a drink when one of the old-timers pulled up a chair.

"Sverre was telling me you hugged him before you leave."

"Yeah."

"You wanna hear something funny? This is what your dad said: 'Yeah, my boy Edgar, he hugs me before he leaves for Alaska. I don't know what to do, but I sure do like it.'"

Speaking of Edgar, the last time we checked in with his saga, he was a long-haired punk blowing all his crab-fishing money on the fast life. That couldn't last forever, and it didn't. In 1995 Edgar was twenty-four. Mom had been telling him for months that there was this nice Karmoy girl who had moved here for a year to live with her brother, to get away from home and check out her options. She was nineteen.

"Why don't you go show her around?" Mom said. "She doesn't know anybody. She's got a real nice personality."

Edgar thought, *We all know what a nice personality means: three hundred pounds plus.*

"It's OK, Mom," he said. "You don't have to set me up with anyone. I'm doing all right on my own."

He'll never forget the next day. It was May 17, Norwegian Constitution Day. He was downstairs sleeping on Mom's couch. She woke him up and said, "The Karmoy club is having a *syttende mai* dinner. You want to go?"

"I don't know."

It took her a few minutes to talk him into it, but he ended up going down. Edgar remembers, "I was looking across the table and there's this gorgeous blonde. She's eyeballing me, too."

The girl's brother called Edgar over and introduced Louise.

"I kind of stared at her," Edgar remembers. "We met, smiled, and did a little chitchat. Then we sat down for dinner across the table from each other. We were speaking mostly English, but a little Norwegian, too. She had a really thick accent. By now we'd each figured out who the other person was. Then we went out to the hallway for a cigarette, and talked. Turned out she grew up two blocks from my mom's house. She had to go to the airport that night to pick up a friend who was flying in.

"Well, since you're from Norway," he said, "do you know where the airport is?"

"Sort of," she said, smiling.

"I'll drive you down," he said.

"Really? OK."

That's where it all started. They hit it off from there. They've been together ever since, and have two sons, Erik and Logan, and a daughter, Stephanie. They were married at Aakra Kirke in Karmoy, in a dual wedding with Louise's identical twin sister. It was the same church where our parents were confirmed—a fitting place for the ceremony. "I tell my kids a hundred times a day, I love you," says Edgar. "I hug them and kiss them in front of their friends. Because I do love them, and I'll tell them straight up."

One strand of the Hansen saga that remains to be tied up is that of my brother Norman. As far as the family business, he was out of the picture for fifteen years. We thought maybe he'd never come back. Yet Norman is just as much a Norseman as any of us, and something about the idea of him living hundreds of miles from the sea, working on cars instead of boats, just didn't sit right.

At Edgar's wedding, Norman met a girl from Karmoy, and they started dating. To travel back and forth to see her, he needed more money than he was making as a mechanic. So about five years ago, Norman decided to get back into fishing. We welcomed him back on the crew. He jumped back on the hydraulics like he'd never missed a day. In the time he'd been away we had had the control panels redesigned, but for him it was like riding a bike. A lot of guys are jerky on the controls and whip the handles around, but Norm is as smooth as silk. It's good to have him back.

Norman still lives out in the forest in eastern Washington. He's not a big spender. He lives simply, and drives a small Toyota pickup. His earnings sit in the bank. His accountant gets mad because it's not accruing a lot of interest. Norman has never even had a credit card. Everything he does is with cash.

"I'll keep doing it as long as the body holds up," he says. "I've still got all my tools from the dealership, so if I have to fall back on mechanics, I will."

Norman is as steady as they come. "I've got full faith in our boat. We got nailed pretty hard a few times. Went almost ninety degrees. But I was sure it would bounce back. If you know your boat, and have faith in it, you have nothing to be afraid of. Nothing scares me," Norman says. "Except spiders. And shopping."

Norman's a bit perplexed with the fame, including the hundreds of fans who've added him on MySpace. "I've got one thousand seven hundred friends I don't know. They watch the show and send a lot of notes and questions. I find it flattering."

One problem with going back to crab fishing is that it interferes with the autumn hunting season. "The only thing I hunt now are gray diggers—squirrels that will destroy your property, tunneling everywhere. So I leave a twenty-two by my door, and if I see one I try to shoot it. Eighty percent of the time I miss."

He's also inherited a flock of wild turkeys. The previous owner of his property introduced fifty turkeys ten years ago. Part of the deal when he bought the place was that he keep feeding them each winter. He goes to the Department of Fish and Game, and they give him free turkey feed. Hunting the turkeys is forbidden.

"I have a turkey feeder up there," says Norman. "At last count I had fifty-six, so they're breeding. They're getting bigger and bigger. I keep all the turkeys fed. I see hunters out looking for them, but all I have to do is drive up my driveway and I'll see twenty of them. It's kind of like a sanctuary. They know they're safe."

Looking back, I still don't know if I became captain of the *Northwestern* through destiny or ambition. When I was nineteen, Dad flew up to relieve Tormod for a trip. Back then the seasons were so long that we didn't have to push it as hard as we do now.

"We'll go out when we go out," Dad said. "It's no big deal." But it was a big deal. We ran around Dutch to buy supplies and food to get ready. I was very excitable and gung ho, and I wasn't willing to be late because of Dad's old-time attitude. I didn't want to spend an extra minute in Dutch Harbor. The crew wanted to fly in and get to Akutan as soon as possible to pick up our pots and bait.

"Ah, the season's gonna go on for months," said the Old Man. He would have been happy to spend a couple days in Dutch and see all his old friends. "What's the big deal? Take a day off. Take a break."

I knew better. The night before our scheduled pickup in Akutan, we had stayed out late in Dutch Harbor. Early in the morning, I was the first guy up, rearing to go. Dad was still asleep in his stateroom. Opening Day was approaching, and Dad was sleeping late.

I was nineteen, one of the youngest guys on the boat. The other men were much older and had known my dad for years. I looked at them. They looked at me.

"Throw the lines," I said.

I went up to the wheelhouse and fired up the engine. I knew how to semi-find my way around the charts, and semi-navigate, but I'd never run the boat before. The men did what I told them, and we steamed out of Dutch Harbor.

Right about when we got to Akutan, Dad woke up. He rubbed his eyes and looked around. He was confused.

"We're almost there," I told him. He mumbled something, unsure of where he was or what was happening. He wanted a cup of coffee.

"Dad," I told him, "it's time to go fishing."

I had shanghaied my own father.

10

DELIVERANCE

Captain Sverre cocked his head. He was slumped and soggy in their freezing floating waterbed. He heard something, and then the others heard it, too. Were they hallucinating? It sounded like the growl of a diesel. They pulled back the flap, but didn't see anything.

"Give me a paddle," Sverre said. He wanted to spin the boat so he could look in the opposite direction. Just then they heard a voice. "Are you alive in there?"

Sverre spun the boat and there through the flap was a miracle. Looming in the mist was the sheer flank of the *Viking Queen* out of Petersburg, another crabber from the Dutch Harbor fleet. The skipper was Joe Lewis.

"We saw the smoke," called a deckhand, "so we figured someone was in trouble."

One by one Sverre and his men were hoisted out of the raft and onto the deck of their rescue boat. He'd survived! It was over. He wanted to rush up to the wheelhouse and thank the skipper, but when the deckhands released their grip on his shoulders, Sverre's frozen legs buckled beneath him, and he collapsed to the deck. He rubbed his legs. They would be all right.

The crew put the men in a hot shower, and gave them the few extra warm clothes they had. One of the crewmen was a huge Hawaiian. His jacket was so big that Leif and Sverre fit in it together. They steamed to Dutch Harbor and delivered the men to safety.

Sverre gave the skipper permission to pull all the *Foremost*'s pots and help himself to whatever was inside. The *Viking Queen* did damn well—picked up a half a million pounds before Christmas. Even in defeat, Sverre knew where the crab were.

My family still visits Karmoy all the time. We bought the house my mom grew up in, and Uncle Karl bought the house that he and my dad grew up in, so we're still very connected to the town. My mom's mother, Nelli Jakobsen, is still alive. She turned 102 years old last summer. She is the last of a generation in which the women worked in the fish plants while their husbands worked on the boats. Draped in a rubber apron, she salted the herring in huge barrels for many years. There is not much fishing industry there anymore. In the past few decades, Norway struck it rich with oil exploration, and now according to the World Bank, just sixty years after the ravages of war, it's the second richest country in the world, after Luxemburg. Karmoy now has waterfront condos, tourists, and small pleasure boats.

Over the years I had had a few girlfriends, but none of the relationships lasted. Then I became friends with a woman named June Kvilhaugsvik, daughter of a Karmoy fisherman. She was born in New Bedford, but her family moved back to Norway eighteen months later, and she grew up down the street from Mom's house in Karmoy. June (pronounced *yoo-na*) had married a Karmoy fishermen when she was eighteen. When we met, she and her husband and their daughters Nina and Mandy were living in Seattle. After six years in the States, the couple moved back to Norway and soon divorced. June and the girls had to start over. June took a job as a manager of a clothing store, and stayed in Karmoy the next seven years. During my regular visits to my family in Karmoy, June and I got to know each other better. Now and then we talked on the phone. I began to trust her more than anyone. I felt like she truly understood me. Her family had been fishing for three generations. Her father, Njal Kvilhaugsvik, fished herring in Norway, scallops in New Bedford, and halibut in Alaska, and June understood the life that fishermen lead. She got it. The phone calls became more frequent. I saw her when I visited Karmoy.

Finally I invited June to come back to Seattle for a visit. About that time my parents were visiting Karmoy, and they got together with June's parents, who, of course, they'd known all their lives. I can only imagine what was said, but the next thing I knew, the Old Man had bought June and me plane tickets for a weekend in Las Vegas. So we went—and that was our first date. I guess it's pretty strange to have your Old Man set you up—and stranger still that he'd send us to Vegas. But no matter. I knew that this was the woman I wanted to spend my life with. We were married just a year and a half later—at Aakra Kirke church in Karmoy, naturally. Things have a way of working out. It was meant to be.

Dutch Harbor is still an outpost, but now has a lot of money. Even after the crab industry flattened out, pollock and cod fishing took off, and these days make as much as three hundred million dollars per year. Everyone has cell phones and the days of waiting for the phone booth in a blizzard are over. The roads have been paved. Carl's Hotel and the Elbow Room closed down. Now there's a luxury hotel, the Grand Aleutian, with a white-tablecloth restaurant serving "North Pacific Rim Cuisine." When they built it, I thought they were nuts, but I underestimated Dutch Harbor. There was plenty of money to keep it afloat.

The crab fleet is dwindling. By 2005, the final year of the Derby, the number of boats had risen to 250, as high as it was during the golden era. That ended the next year when the Derby was replaced by IFQ—individual fishing quota. Based on past performance, each boat was assigned a certain amount of crab, and could catch it at their own pace.

As a result, the industry is no longer open to everyone. If you don't have quota, you have to lease or buy it from someone who already owns it. Many of the smaller operations that got quota decided to stay on dry land and let someone else do the work and take the risks. They leased their quota to the bigger ships and got out of the business. The fleet shrank from 251 to 89. Five years later there are fewer than 50 active vessels.

In some ways, IFQ has made it safer, because we have the option to stop and rest. But do we still go full bore? Absolutely. We want to get our crab caught before the ice moves down from the Arctic. We want to get it to market when the market dictates. Time is still money, and it costs a lot to run that boat. So if it takes two weeks to catch what could be caught in one week, we've wasted ten grand on fuel.

We're like athletes. Once the gun goes off, we want to win. That's been bred into us for generations. It's in our genes. Even if we are fishing side by side with our best friend, we're still going to lie to him, try to do better than him. If you can tell him that you got fifty thousand more than him in the same amount of time, you can tell him you're the better fisherman.

In the early nineties, Dad's years of hard living were taking their toll. He had become diabetic, and took daily meds. Try as he might to get healthy, his body was failing him. On one trip up the Inside Passage to Alaska, he felt so ill that Edgar dropped him off in Ketchikan to fly home.

When Edgar's son Erik was born, Sverre became a doting grandfather. He became best friends with the boy. With Edgar's wife, Louise, working full time, Sverre and Snefryd had regular babysitting duties. Sverre toted the toddler along the docks, the same docks he had walked in search of work forty years earlier. He bought the boy ice cream and spoke to him in Norwegian. Erik loved his gramps.

One time Sverre took Erik to Norway. "I went first class," he later told me, almost in a whisper. "Don't tell anyone." That was a big deal. My parents were never the type of people to flaunt their wealth. They did have a big house for their time, and he did have a big boat of which he was very proud, but they didn't boast. They didn't have massive parties. They never went to the yacht club. They were very low-key.

Sverre also doted on his nephew's children. When Stan had a daughter, Sverre would drop by unannounced to visit the little girl. The grandchildren provided another bond between Dad and his brother.

Sverre was overjoyed when Edgar and Louise had a daughter, Stephanie. Her baptism was planned for Sunday morning, June 10, 2001. I was in Norway with June, unable to make it. When I spoke to Dad on the phone, he told me he wasn't feeling well, and was going to see a doctor. He had just been in Norway and June's father had given him a tub of herring. Sverre had salted and brined it, and eating it was giving him heartburn. He made a doctor's appointment for Monday.

At the time he was so concerned with his health that he was keeping a daily journal of how he felt and of the meds he was taking. I told him I would come home and he told me not to. Sverre wanted me to be in Norway with June. He wanted me to get on with my life, and was happy that I was there. Norman was in town, staying at my parents' house, so the baptism would be a nice family event.

When my mother awoke that morning, Sverre wasn't in the room. That particular evening he had fallen asleep in the den in his favorite chair as he watched television. She went to the kitchen to make coffee, then brought it to him. She found him lying on the floor, flat on his back, hands by his side, with a slight smile on his face. "Get up," she said. But he lay still. She thought he was joking. She called his name, suddenly afraid. He was motionless, just like when you turn off a switch. Nothing. She screamed. Norman came running up the stairs, knelt beside our father, and grasped his wrist for a pulse, but knew by the cold skin that he would not find one.

I don't remember the phone call. I don't know how I found out. I was in Norway and somehow the news reached me. Maybe my mom or Norm or Edgar or Uncle Karl called to tell me. They, along with my cousins and Oddvar and his wife, had also converged at the house immediately upon hearing the news that morning, The support was instant, unquestioning. I was numb.

June and I boarded the next available plane and four days later I was standing in a suit before a packed congregation. A sorrowful hymn was sung in Norwegian. My mom and my brothers were there in the front row, beside Uncle Karl and Aunt Else and my cousins.

Edgar and I both gave eulogies. "It wasn't always easy," I said. "But nothing good ever comes easy." My dad was not the most affectionate or emotional man. He wasn't one to say I love you. He wasn't one to get choked up or shower you with praise. Mom told me later that he paced the floors worrying the whole time I was gone fishing, but I never would have guessed it at the time.

I spoke about a time when I was twelve years old, the first time he took me up on the *Northwestern.* We went up the Inside Passage, and when we crossed the Gulf, the weather was terrible. I was so seasick. Not only was I sick all day, and not only did I lack the strength to get out of my bunk, but the seas were so rough that I was getting tossed off the bed and onto the floor, barely able to crawl back up. I was miserable. This went on for days. My father was up in the wheelhouse—he could stay on watch for days with no sleep. Then he came down to my stateroom. He had an old dirty piece of plywood. The thing stank. I don't know where he got it. He jammed it between my mattress and the edge of the bunk, so it formed a barrier. Now instead of falling on the floor, I'd bounce between the wall and the plywood. It might not sound great, but it was a lifesaver. That day, I knew he really cared for me. I never even expected him to come down from the wheelhouse. As a twelve-year-old kid, I thought it was the nicest thing he could do for me.

My father didn't spoil us. He taught us to be as humble as he was. He taught us to have pride in our family, as he did. He showed us that we should give, and be grateful for how fortunate we are. Our family has been blessed to have had such a wonderful father. He truly loved and respected my mother. He showed us what it meant to be a truly good person. He was a good father to me, and he became my best friend.

Even as I spoke these words at his funeral, I felt somewhat numb. I didn't weep. I kept it together. I don't really remember what I felt.

I guess the grief didn't come until later. The next week, I was back in Alaska. Things were going well for me. We were making money.

My brothers and I would become partners in the family business, equal owners of the *Northwestern*. June and I would be getting married soon, and she and Nina and Mandy would join me in Seattle. Eventually I would buy my parents' house from my mother and raise my family where my father raised us.

Then one night I was taking the *Northwestern* out of harbor, into the Bering Sea. Edgar and the crew were asleep in their bunks. The sky was black and the sea was ink. All that was visible was the jagged outline of the mountains. The boat rose and dropped in the waves, and sometimes a plume of spray showered the wheelhouse. A flock of white gulls bobbed on the water. As the ship bore down they lifted in unison, beat their wings, bared white bellies to the ship lights as they peeled off and soared in the wind.

Suddenly I lost it. I doubled over, exploded in tears, and bawled like a baby. As soon as I sat up and shook it off, the grief racked my body again. I sobbed. All that I had held in during the funeral and the past weeks erupted. I had no one to look to for guidance. It was up to me to deliver my family to safety.

My father was gone, the ship was mine, and I was alone at the wheel.

You won't hear Norwegian spoken on the streets of Ballard anymore. It's become one of the trendiest neighborhoods in Seattle. The old flea bag, the Princess Hotel, next door to the Smoke Shop, has been converted into the upscale Princess Apartments. Downstairs is the slick office of a public relations firm. On Saturday mornings Ballard Avenue is blocked off for a farmers' market frequented by sandal-wearing recyclers. The storefronts that once sold hardware and ship supplies are now boutiques called Dolce Vita, Ella Mon, Asher Anson, and Came-

(Courtesy of EVOL)

lion Designs. It's not clear to me what they sell. Instead of a shoe store
we have Kick It Boots and Stompwear. The Norsemen, Scandie's, and
the Troll Café are gone, replaced by Volterra, Thai Ku, India Bistro,
and the Matador Tequila Bar. What New Ballard really has to offer
is coffee shops. On Market Street there are three in a row: Starbucks,
Tully's, and Verite. If you can't find enough options there, just cross
the street to the Chai House. If you're still not satisfied, there's Floating Leaves Tea one block away on Ballard Avenue, next to the site of
the old Ballard Tavern.

One of the few holdouts from Dad's day is Johnsen's Scandinavian
Foods, which changed its name to Olsen's a while back. These days
Norwegians don't come in day-to-day like when they lived in the
neighborhood. Maybe a granny from Olympia or Edmonds will come
in, usually before Christmas or May 17, to prepare for a special feast.
Sometimes a young professional from the neighborhood will hold a

Norse-themed dinner party and need instructions for fishcakes. The store makes much of its income from online orders. At Olsen's you can still get salt cod and dried lamb sausage. A few doors down is something called Shakti Vinyasa Yoga. I'm not sure what you can get there—I've never been in.

A few other remnants of Old Ballard remain, if you know where to look. Each May 17, there's still a big parade. Edgar and I and our families march in it. My wife and daughters don the traditional Norwegian dresses, and we carry Norwegian and American flags as we walk down the avenue. It's a great reminder of the way things were. Old-timers meet daily at the Leif Erikson Lodge in the kaffestua—coffee room. An old Swede plays the accordion while the white-haired crowd lunches on open-faced sandwiches of pickled herring and deviled eggs, mini heart-shaped waffles, and bowls of thick homemade potato soup.

The old guys aren't pleased with the neighborhood changes. In 2001, when the Port of Seattle remodeled Fishermen's Terminal, it broke with decades of tradition and allowed pleasure craft to berth beside commercial vessels. In the last few years, developers tore down the Denny's and the Sunset Lanes bowling alley, places that local kids had hung around for generations. More condos were coming. The only problem is, when the economy collapsed, the investors bailed. Now there are gaping vacant lots.

The ultimate outrage came when the trees in Bergen Square, planted by King Olav himself, were cut down and replaced by "Witness Trees," apparently a modern art masterpiece. Instead of fir trees, we now have an Immigrant Tree, a Clam Tree, and a Fossil Tree—old wooden power poles capped in stained glass, wrought iron, seashells, and ceramics. "Drug mushrooms" is what the old-timers at Kaffestua call them. "If they want art," griped one, "they should put it downtown in the museum."

Down by the waterfront, there's still a bit of the old Ballard—at least for now. It's an industrial zone of derelict trucks and train cars, a bit of graffiti, Dumpsters, crumbling brick and concrete, and silos of fuel. Marco shipyard is still standing, but barely, an abandoned husk of corrugated steel surrounded by slick new buildings. A sign on its flank reads OFFICE/CONDO AVAILABLE. Pacific Fishermen is still in business—at least the fleet still has an active shipyard. Jacobsen Marine recently shuttered and moved, but Lunde Marine Electrical is still open, owned by Tor Tollessen, who grew up next door to my dad in Karmoy, and whose father-in-law was Inky, owner of the long-gone Ballard Tavern.

Of the five marine fuel stations, only two remain. Ballard Oil is still run by Warren Aakervik, a Norwegian-American who bought the business from his dad in the eighties. It hasn't changed much since 1937, when it was founded. A local legend with his grease-stained overalls and walrus moustache, Warren has been the master salmon-griller at the yearly Ballard SeafoodFest for thirty-three years and counting.

These days he's pessimistic about the future of the maritime industry and Ballard Oil—under fire from all directions in this boutique neighborhood. The bureaucrats in the environmental agencies require twenty-four-hour written notice of nearly every fueling that occurs in his dock. The pencil pushers at Homeland Security have forced him to build tall barbed-wire fences and restrict dock access to only licensed personnel. The City of Seattle wants to build a bike path along the back of his property, where his eighteen-wheel fuel tankers come and go. A single accident could cause a spike in liability insurance and drive him out of business. At seventy-one years old, he's not sure now much longer he wants to continue. He and his family have run a safe and responsible business on Salmon Bay for seven decades, one that adds wealth and character to the community. If the bureaucrats have their way, fishermen will get their fuel elsewhere, and Ballard will have little more than lattes and memories of its days as the hub of the North Pacific fishing fleet.

As for the crew of the wooden *Foremost*, the last time I saw Magne Berg I was a teenager, and I remember him stumbling off a gang-plank, a hell-raiser to the end. He moved back to New Bedford where his parents lived, and he died in his fifties. Krist Leknes moved up to Whidbey Island in Puget Sound, no wife or kids. Some say he moved up there to get away from all the Karmoyverians. He died there. Leif Hagen still lives in Seattle, a widower with no children, in a house right down the street from his best friend and my first skipper, John Jakobsen. He's the last survivor of the *Foremost* disaster.

John Johannessen and Borge Mannes passed away a few years ago. Oddvar Medhaug and Tormod Kristensen moved back to Karmoy, and just as I was finishing this book, I got the sad news that Oddvar, one of my greatest mentors, had passed away. Magne Nes is still fishing at age seventy-six—building and experimenting with wooden tuna boats in California. Jan Jastad and John Jakobsen and Pete Haugen and Gunn-leiv Loklingholm are still going strong. My uncle Karl survived heart surgery, still smokes two packs of Winstons a day, and still lives in the neighborhood. I bring him herring by the crate, and he cures it in five-gallon buckets in his garage. In exchange, he cold-smokes my salmon that I bring home from Alaska. It's a delicacy. I'll go through half a side of that stuff just sitting on the couch watching a movie.

The Karmoy boys, old men now, meet almost every morning at a Ballard diner, drinking coffee and swapping stories, the same ones they've been telling since I was a kid. They're a bit of living history. Nearby is the towering statue of the fishermen's memorial, where my great-uncle Jorgen Hansen is inscribed on the bronze wall, along with names like Jacobsen and Eriksen, Haagensen and Carlsen, Petersen and

Christiansen, Larsen and Torgramsen. Hundreds of lost sons. Among the plaques on the walk is one that commemorates my dad's life in the fishing industry.

I suppose I'd like to keep things the way they were, the way my father kept them. After we renovated the boat last year, I came on-board the day we were to launch and found that the galley picture—the one of the bearded old man praying before his meal—had been removed so the workers could refinish the wall. I wouldn't leave the harbor until I got it back. We found the picture and restored it to its rightful position—same holes, same screws, same place. Then we were safe to leave port.

One place in Ballard that remains unchanged is the Smoke Shop. Newspaper boxes line the curb, but the *Post-Intelligencer* kiosk is empty since the paper stopped printing. A sign in the window advertises the day's special: MEATLOAF WITH MASHED POTATOES AND GRAVY, CARROTS, SOUP OR SALAD, ROLL AND BUTTER, $7.25. Below that hangs a flyer for a $2.50 bag sale at the thrift store next door, and next to that a stern warning, NO PUBLIC RESTROOMS.

But come inside. At night you'll get a mixed crowd, young kids and hipsters spilling over from the trendy nightclubs. If you hit it right, in the morning, it will be just be the regulars. The afternoon barmaids Marcia and Darlene and Barbara have worked there for decades. Framed photos of fishing vessels still line the walls. The *Northwestern* is up there, and so is Karl's *Ocean Spray*.

You can't smoke in the Smoke Shop anymore, but there's still a ciga-rette machine, for old times' sake. The regulars sneak past an EMPLOYEES ONLY door through a storeroom to the alley staircase to light up. The jukebox still plays "North to Alaska" and "Sink the *Bismark*." The old steel bell hangs over the bar, not rung as much these days as when the highliners used to steam in from Alaska loaded with crab and cash.

Tom Economou is there at six most mornings. On Fridays his kids come in for lunch. His older brother and partner, Pete, passed away a few years back after three decades of running the restaurant. As sole owner, approaching his eightieth year, Tom has respectfully kept the place as it was. He gets offers to sell the Smoke Shop all the time, but he turns them down. "I grew up here," he says. "Thirty-eight years. I'd get a bunch of money—then what?"

So he's there at six, serving stiff drinks for the regulars. See for yourself. Some white-haired sailors may be stooped over the bar, sharing stories, and if you lean close you'll hear their Norse accents. Maybe one will nod at the wall at one of those shimmering steel crabbers, the *Aleutian Spray* or the *Nordic Viking*. Listen closely, and he'll tell you his saga.

Acknowledgments

*I*nterviews with Dick Pace, Lloyd Cannon, Bart Eaton, Ole Hendricks, Lowell Wakefield, Joe Kurtz, Cap Thomsen, Peter Schmidt, and Ed Shields are from the excellent documentary *Pots of Gold*, by Bob Thorstenson and writer-producer John Sabella in conjunction with the Nordic Heritage Museum.

Thanks to all who were interviewed: Snefryd Hansen, Edgar Hansen, Norman Hansen, June Hansen, Karl Johan Hansen, Else Hansen, Stan Hansen, Leif Hagen, John and Darleen Jakobsen, Jan Jastad, Oddvar Medhaug, Lloyd Johannessen, Magne Nes, Tormod Kristensen, Charlie McGlashan, Howard Carlough, Jan Andersson, Doug Dixon, Buddy Bernstein, Mangor Ferkingstad, Pete Haugen, Marvin Hanson, Sigmund Andreasson, Tom Economou, Nick Mavar, Jr., Matt Bradley, Jake Anderson, Chris Aris, Mark Peterson, Mike McCool, John Sjong, Knut Thorkildsen, Larry Hendricks, Joe Sanford, Tim Canny, Tor Tollessen, Marsha Stiff, and Soren Sorensen.

I want to thank my agent Richard Abate for sending this project my way, and also Shawn Coyne, Dorian Karchmar, Peter Wolverton, and Elizabeth Byrne for carrying it to completion. I am grateful for Brian Kevin and Andy Smetanka, whose research and editorial assistance was prompt, professional, and precise. I would like to thank Sharon Oard Warner and Greg Martin at the University of New Mexico and the Taos Writers Conference for offering me good places to work and write. And I am indebted to my colleagues and students at the West-Conn MFA in Creative and Professional Writing Program, including Brian Clements, Laurel Richards, Don Snyder, Daniel Asa Rose, Elizabeth Cohen, Paola Corso, and Dan Pope.

—MARK SUNDEEN

INDEX

Hansen, Norman (brother)
  background of, 65–67
  begins fishing, 65–66
  early life of, 12–13, 34–37, 65–67, *164,*
    227–28
  education of, 66–67
  love of hunting, 67, 287–88
  on the *Northwestern,* 21, 25, *33,* 65–66, *129,*
    132, 286–87
  Peterson and, 34, 78, 79
  sibling rivalry with, 33–36
  on television, 33–34
Hansen, Sigurd (grandfather), 11–13, *41,*
    42–43, 60, 223–24
Hansen, Snefryd Jakobsen (mother), 108,
    189–90
  christening of *Foremost,* 223–24
  christening of *Northwestern,* 238, *239,*
    239–40
  courtship and marriage to Sverre,
    137–38, 140, 142
  death of Sverre, 296–97
  in Dutch Harbor, 207–8
  family life, 12–13, 26, 36–37, 227–28,
    285–86
Hansen, Stephanie (niece), 295–96
Hansen, Sverre (father)
  advice from, 15–16, 18, 19, 26, 154–55,
    156
  Aleutian Monarch and, 244–45, 251–52
  in the Army, *105,* 105–6, 136–37, *137*
  arrival in Ballard, 64–65, *65,* 90,
    99–100
  begins fishing, 62–64
  cleanliness of, 80, 283–84
  courtship and marriage, 137–38, 140,
    142
  death of, 295–98
  early life in Norway, 4, 56–62
  Edgar and, 69–73, 284–85
  education and, 27–28, 62, 89
  Enterprise and, 252–53
  family life, 33–36, 207–8, 227–28, 295

on the *Foremost* (steel), 223–26, 231
on the *Foremost* (wooden), 45–48, 51–56,
    222
  escape on life raft, 145–46, 179–81,
    209–10, 241–43, 263–64
  fire onboard, 1–8, 75–77, 111–13
  repairs, 45, 46–47
  rescue by *Viking Queen,* 291–92
  sinking of, 209–10
living up to the example of, 9–10
love of music, 104–5
love of Norwegian food, 206–7
McGlashan and, 141–42
Norman and, 66–67
on the *Northwestern,* 13–15, 20–22, 43–44,
    255–62, *256,* 282–85, 297
  christening, 238–39, *239*
  giving authority to Sig, 148, 154–55,
    156, 288–89
  launch day, 239–40
  lengthening of hull, 247–48, 249,
    252–53
  purchase, 231–32, 238
partying of, 117–19, 138
on the *Seattle,* 140–41
sibling rivalry, 32–33, 138–39, 228
successes of, 205–6, 247–49
Tormod and, 80–81
U.S. citizenship of, 208
on the *Western Flyer,* 4, 100–105, *102,*
    137–38, 142–43, 188–90, 201–3
Hanson, Ole, 97
Harald Fairhair, 39
Haro, Gonzalo López de, 198–99
Hattie's Hat, 206
Haugen, Arnie, 117
Haugen, Pete, 65–66, 117, 302
Havana, 121
Havbell, 114–17
Hazing, 18–19, 25–26, 81–83
Helly Hansen, 276
Hendricks, Ole, 175–76, 250
Hera, 161

Union, Lake, 97–98

Unions, 103–4

UniSea (Universal Seafoods), 173, 244, 246–47, 248

University of Oregon, 118

University of Washington, 171–72

Vancouver, George, 197

Vasa, 119

Vasa Sea Grill, 119–20

Vassvaag's sausage shop, 60–62

Vessel Safety Act of 1988, 253–54

Vigra, 42, 63

Viking Queen, 291–92

Vikings, 28–32, 38–40, 233

Vinland, 31–32

Vita Seafood Products, 244

Wakefield, Lowell, 171–73

Wakefield Seafood Company, 45–46, 47–48, 52, 172–73

Walker, Spike, 246

Walmart, 273, 275–76

War of 1812, 104

Washington, Lake, 97–98

Waves, rogue, 6, 183–87

Weekly Pacific Tribune, 96

Western Flyer, 4, 100–105, *102,* 117, 121, 137–38, 142–43, 188–90, 201–3

Western Tugboat Company, 99

West Ness, 138–40

Westpoint, 234

Wetsuits, 156–58

Wheelhouse, 169–70

Wheel watch, 81

Whitemore, Bruce, 233

Wild Horse, 279–81

Wild turkeys, 287–88

Wilse, A. B., 92–93, 95–96, 97

Worker strikes, 47–48, 228–29

World Cup, 227

World War II, 58–59, 98–99, 104, 141

Xanadu, 245, 252

Yggdrasil, 39

Yost, Gary, 272

Zhemchug Canyon, 194–95

Zooplankton, 165